A NEW MODEL

A NEW MODEL

PONDERING THE REALITY OF GOD

Patricia Hylton Zell

Central Ohio Renaissance Productions LLC

First Printing, 2022

ISBN 978-0-578-35492-7 paperback

ISBN 978-0-578-35493-4 e-book

Central Ohio Renaissance Productions LLC
CenOhRenProLLC@yahoo.com

In memory of my husband, Merlyn Keith Zell, who encouraged me in my quests to understand God, to understand the Bible, and to sharpen my writing skills. His support meant the world to me.

ACKNOWLEDGEMENTS

This book came into being with the help of IngramSpark (especially their book-building tool) and Grammarly Premium. These websites are excellent assets for authors everywhere. I give my heartfelt thanks to the creators and staff of both companies.

The cover photo is courtesy of tombud at Pixabay. Again, I extend a huge thank you to both the photographer and the website.

My greatest blessings are my children (and in-laws) who inspire me daily: John, Liz, Becky (Dave), Jenny, Will (Beth), Lily (Shawn), and Meg. They bring much joy to my life. And their children have added to that joy: Mara (Dathan), Lauren, Elisha, Andrew, Ethan, Mia, Brant, Bella, Grace, Jack, Dom, Zach, and Luke. Then, we've all been blessed with Pam, George, Keller, and Ali, who have rounded out our family.

A special thank-you goes to Goldene, my late husband's sister, who has been my special cheerleader throughout the process of writing this book. Also. a call out goes to my siblings, Tim, Bob, and Sue, who, along with our late brother Dave, were part of the home that shaped me. And, finally, I honor my late parents, Bill and Ilene, who laid the foundations of my life with their focus on getting a good education and paying attention to what was on the inside of me and to my actions ("Pretty is is pretty does").

Contents

1

INTRODUCTION

Keywords: Model, God, Christ, Bible, Science, Perspective, Puzzle Pieces, Chapter Title, Keywords, Chapter Notes, Remember, Reading List, KJV, NASB, Biography, Disclaimer

Introduction to The Model

(1) The work on this new model began when I was a freshman in college, way back in 1969-70. After getting hold of a *Good News for Modern Man New Testament* that year, I decided to read the Bible independently with no Bible study or devotional guide involved. As I worked through the book of Matthew, I started asking God questions about what I was reading. I remember my first one about Matthew 24:15: "What is this 'Awful Horror' (aka 'The Abomination of Desolation') that Christ talked about happening in the last days?" [Many years passed before I understood enough to figure out the answer.] My quest to comprehend God and the Bible began with that question. Within the first decade of my work, I decided that I wanted my understanding to come from God and not from other people. I didn't go out into the desert, sit on a rock, and ask God to tell me everything. Instead, I began

going into my prayer closet with my Bibles, a concordance, a dictionary, and writing supplies to spend time alone with God. This pursuit hasn't been an easy one. The Bible is a complex literary work by at least forty writers whose lives spanned approximately 1500 years. With the figurative language and the elements of historical writings within its pages, the Bible has a depth of meaning beyond any simple, superficial renderings. Thus, especially in the earlier decades of my quest, there were times of intense frustration on my part when things just wouldn't logically come together. I didn't give up but kept pushing God to help me understand.

(2) Around the beginning of the twenty-first century, I began researching scientific discoveries, which helped immensely with my quest. Despite some popular opinions, the Bible and Science aren't at odds. I've come to consider them to be two sides of the coin of reality. For instance, one significant similarity that the Bible and Science share is the multitude of voices involved in each endeavor—at least forty writers for the Bible (consisting of 1000 plus pages) and countless scientists for the worldwide systematic endeavor of Science. A second significant similarity between the Bible and Science is what each presents is progressive. While this reality is apparent in Science, as far as the Bible goes, many people try to give every bit of information in the book an equal weight of importance. Some people consider the Ten Commandments equivalent to Christ's teaching, even though Christ put God's "law" in terms of love rather than a list of rules and regulations. We need to remember that Christ knew God in more advanced terms than Moses. Moses was all about rules, punishments, and appeasing God through various rituals, including the sacrifice of animals. Christ was all about God being our Father and about the importance of how we treat each other.

(3) One misunderstanding about the Bible has grown out of the practice of giving all of its information comparable importance. Some Christians believe that the Bible is infallible. The Bible itself contradicts

this doctrine. In I Corinthians 13:9-12 NASB, Paul says that they who were believers (including those who wrote the New Testament) only knew in part—they didn't have the whole picture. Anything that is only part of the story cannot be infallible. Let's take this even a step further: Genesis 1 and 2 are about Creation and are very different from each other. Is the Bible contradicting itself? Looking at this divergence, Genesis 1 is the story from God's perspective (reality), while Genesis 2 is from the human viewpoint (supposition). In the first chapter, God created the animal kingdom before He created Adam. The second chapter reverses the order. Adam saw something different than what God saw. When God brought the animals to him to be named, Adam assumed that God created them after He created him. Adam didn't know the reality of the situation—God worked His way through the plant and animal kingdoms before He created humans. [Science supports this version of the story.] The lesson we can learn from this discrepancy is that both God's and human perspectives reside in the pages of the Bible.

(4) So, the Bible has information written from God's perspective and the human perspective. As far as Science goes, humans interpret every bit of evidence and all observations. But, as scientists may agree, the human perspective sometimes raises more questions than answers. However, scientists don't give up but keep pressing to find reality. While many religious people cling to theological writings from many centuries ago, scientists are all about rewriting the textbooks. These scientists want to locate the next new puzzle piece because they know they still have much to learn and figure out. Surprisingly, the Bible holds some of the puzzle pieces of reality. Sometimes the pieces are tiny—a verse, a sentence, or a phrase—but they fill in some of the missing aspects of scientific discoveries. When we put the scientific and biblical puzzle pieces together, the information begins to make sense and meld into an understanding that might work. Of course, only the passage of time will reveal whether this knowledge is close to reality or not, which is why I've titled this book *A New Model: Pondering the*

Reality of God. [Just a note here: to make the model easier to read, I'll minimize using words like "perhaps" or "maybe." Please understand that this information is a model, not a certainty.]

(5) Needless to say, I've used a lot of resources in my quest to understand reality. I've handled the Bible without the help of others, but the scientific information has come from work of many scientists. [I don't have the ability or the equipment necessary to do the required research.] I've almost worn out my search engine with multitudes of internet sources as I've tried to put the puzzle of reality together. In light of the copyright laws here in the USA, I'll share only the general discoveries and the facts that scientists have worked out. I won't be quoting any specific scientist/author (except Nikola Tesla on the back cover) or paraphrasing their words. However, I'll include a suggested reading list at the end of most chapters so that readers can peruse the material for themselves. Also, to keep everything organized, I'll set up each chapter in the following manner:

- *Chapter Title*
- *Keywords*: Replacing an index are the keywords in the chapter to make cross-referencing easier.
- *The Model*: I'll number all paragraphs because I occasionally refer to what we discussed earlier in the book. The paragraph labels will make those instances easier to work through.
- *Chapter Notes*: I'll use superscript numerals (1, 2, 3, etc.) in the text to note different scripture verses and passages to make the model easier to read. This section will include the book, chapter, verse(s), and version for each reference (example: [1]John 3:16 KJV). For passages that I quote within the text, I'll include the information in the text (example: In John 3:16 KJV, Christ said ...) or in the Chapter Notes. I'll also use * and ** for other notes.
- *Remember*: When I refer back to previous chapters in the text, I'll use superscript letters (a, b, c, etc.) to indicate the existence of an

earlier discussion. Then, I'll provide its chapter and paragraph numbers in this section (Example: [a] Chapter 3, paragraph 4).

- *Mini Essay*: This section is where I personalize the model through reflections, more information, etc. By separating those interactions concerning the model from the actual model, I'll keep it objective.
- *Suggested Reading*: Although I limit the scientific parts of the model to general discoveries and facts, I'll also share information about the articles that I have found in my research. Reading these articles will provide details of the experiments/studies and the people involved.

(6) One last topic before we start with the model: I have used the New American Standard Bible 1995 (NASB) and the King James Version (KJV) in my quest.[*] As it turns out, these two translations are among the most accurate of all the different renditions of the Bible. Both the NASB and the KJV are word-for-word translations; they're more laborious reading at times, but it's worth it to persevere with them. Many of the newest versions are phrase-to-phrase translations or paraphrases; what we gain from the ease of reading is sometimes lost in the inaccuracy of the wording. One example is John 3:16, where both the NASB and the KJV use the terminology, "His only begotten Son." Many other versions use "His only Son" or "His one and only Son." As we'll see, Christ was the only human being physically born as God's Son. However, through the new birth, every human being has the option to become a son of God. [Christ acknowledged this when he named God as our Father in Matthew 6:9 KJV.] So, for this model, I'll use both the NASB and the KJV. At times, I'll use the wording with the most clout, and at other times, I'll mix the versions up for variety.

Chapter Notes

* Important note: when I quote scriptures from either translation, I make one major change. Wherever the phrase "LORD GOD" is used, I remove most of the capitalization and write "Lord God." This makes reading the passages easier.

Remember

None

Mini-Essay

Here's the shortened version of my biography: I graduated from Eastmoor Senior High School (1969, Columbus, Ohio) and the Ohio State University with an elementary education degree (1973). My life with my husband, Merlyn, began during my second year of teaching. After we quickly started our family, I became a stay-at-home mom. Over the years, we had seven children (two sons and five daughters). Along with my duties as a mom, I continued to seek understanding from God. Then, after our sixth child was born, I took my first correspondence writing course, the catalyst to my second quest of becoming a skillful writer. Those pursuits and raising my family were the foci of my work until 2002 when I accepted a high school English teaching position. I then took the coursework I needed to become highly qualified to teach English and went on to earn my master's degree in curriculum, instruction, and assessment. I taught for eleven years and retired at the end of the 2012-2013 school year after my husband had passed away in the fall. [During my teaching years, I still worked on my two original quests.] Since I retired, I have learned how to write screenplays and increased my time researching current scientific discoveries. After I finish my work on this book, I will return to developing my scripts.

Now, for the disclaimers: 1.) Although this model talks about immortality, I do not promise anyone immortality. I know the responses to this model may include, "But, what if you don't attain immortality?" Then I won't. However, I will have lived a satisfying life full of love and joy. If there's an iota of chance I can attain immortality, I'm going for it. Why wouldn't I? 2.) I am not involved in any ministry. If anyone wants to label me, I am an explainer and writer. If I understand a concept, I can explain it through the written word. I have no authority in anyone's life (my children are grown), nor do I desire any jurisdiction in anyone's life. This model comes out of my relationship with God. Its purpose is to encourage readers to deepen their relationships with God. Also, I am not a speaker because of a physical quirk. 3.) I will not accept (nor ask anyone for) donations.

2

BEFORE THE BIG BANG

Keywords: Big Bang, Bible, Science, God, Generator, Darkness, Light, Cosmic Rays, Model, Reality, Nuclei, Life, Energy, Holy Spirit, Love, Strings, Vibrations, Sound, Eternity, Infinity, Parallel Universe

The Model

(1) To fully understand the reality of the Universe and how it came into being, we must look at what was present before the Big Bang. Both Bible believers and scientists will benefit from this understanding because it sets the foundation for the existence of the space our Universe occupies. Scientists have been asking questions about the Big Bang and what preceded it. They've offered such theories as cosmic inflation, multitudes of universes, and black hole cosmology (the Universe exists within a black hole). Perhaps, this "big picture" model will make sense and answer some basic questions that have plagued scientists for a long time. And, for the Bible believers, this big picture may go a long way to answer questions about who God is and why evil exists. This model will set a foundation for understanding what Christ accomplished through his resurrection. And hopefully, this model will clear up misconceptions that have kept people from attaining all that is available to all of us.

(2) Before we look at what existed before the Big Bang, there are a few principles that we need to discuss. These basic concepts will facilitate our understanding of what existed before and after the Big Bang happened:

- Darkness is a function of a light source: darkness and light are opposites of each other, yet both of their existences strictly rely on the presence of a light source. Light is the result of a light source being turned on or uncovered. If a light source is turned off or covered, darkness is the result. However, darkness has no origin of its own—we can't turn off darkness and produce light. So, the state of a light source determines whether light or darkness exists. This understanding is vital to comprehending what happened during the Big Bang and afterward.
- Like attracts like: Recently, scientists have shown that two objects with the same (but unequal) charge don't always repel each other. Thus, there is evidence that like does attract like in some circumstances. This concept will come into play to explain what happened during the Big Bang.
- Cosmic rays exist: Perhaps one of the most significant indications of the reality of God is the presence of cosmic rays that continually enter our Earth's atmosphere. Scientists don't know the origins of some of these rays, which carry high-energy subatomic particles and the nuclei of every known element. [God could generate the rays.]
- Time is only relevant here on Earth: Time is strictly a matter of the planet's movement (the rotation on its axis and the revolution around the sun). Our time has no impact anywhere in the Universe except here on Earth. Rather than speak of time, we may want to speak of existence, which is not dependent upon time. Existence, not time, was a significant factor in the Big Bang.

With these understandings in mind, let's think about what existed before the Big Bang.

(3) The first question for us is, "Who is God?"—some think of God in terms of what He does rather than who He is. Let's start by looking at the physical Earth and the Universe where matter exists. Everything involved in the existence of matter had to have a source: the energy, the frequencies, the particles, the strings, and the vibrations that made and held matter together had to come from somewhere or someone. Let's call that source God and look at Him as a self-fueled, self-sufficient generator. In Isaiah 45:7 KJV, the prophet quotes God as saying, "*I form the light and create darkness ...*" (one of the definitions of the verb "to form" is to bring into existence). And 1 John 1:5 KJV says, " ... *God is light, and in him is no darkness.*" God generates light and is its source. [As we'll discuss in the next chapter, God also created darkness.] The light that God generates encompasses the whole electromagnetic spectrum. In Revelation 4, the writer mentioned the seven spirits of God—he told of the seven lamps that burn before God's throne.[1] As it turns out, the electromagnetic spectrum happens to be divided into seven major categories: radio waves, microwaves, infrared waves, visible light waves, ultraviolet waves, x-ray waves, and gamma-ray waves. Voila, the seven spirits equal the seven lamps equal the seven categories of light frequencies that God generates.

(4) But, light frequencies aren't all God generates. Remember the cosmic rays mentioned above? Scientists have documented that 1.) those cosmic rays carry the nuclei of every known element and 2.) some of the rays decay into gamma-rays. We could hypothesize that the cosmic rays are bundles of light energy and nuclei, with the rays breaking apart when they hit the Earth's atmosphere. What if God generates the nuclei along with the gamma-rays? In the Book of Psalms, David made a powerful statement: "*For with thee is the fountain of life: in thy light shall we see light.*"[2] God becomes the fountain of life as He generates the nuclei of all the elements. Those nuclei are the basis of life here on the Earth—without those elements, there would be no life. Also, if the nuclei and light frequencies are bundled together, we can infer that

there would have to be an underlying energy force to keep the cosmic rays moving. The Bible contains many references to the Spirit of God, aka the Holy Spirit, and some of those references connect God's Spirit to movement. [Genesis 1:2, John 3:5-8, and Acts 2:1-4 are examples.] The Hebrew word "Ruach," which is often used in the Old Testament to refer to God's Spirit, is defined as spirit and wind. So, it is reasonable to describe the Holy Spirit as God's energy—the underlying force that causes the other things God generates to move. [Note: the seven Spirits of God (the electromagnetic spectrum) and the Holy Spirit (the underlying energy) are unique entities.]

(5) What else might God generate? Bible writers often indicated that God spoke out loud to and through people. Since then, some have ridiculed the notion of God speaking to humans, but what if there's a possible mechanical explanation that supports the occurrence of God talking to people? Let's go back to Revelation 4:5 KJV, where John described a vision he had of the throne of God. John wrote that lightning, thundering, and voices proceeded from the throne. Thunderings and voices are vibrations that manifest as sound waves. Not only was tremendous energy coming from the throne of God, but sound waves also were coming from there. It makes sense that vibrations would accompany all of the light, the nuclei, and the energy that God generates. God's sound waves could have meaning if there was a process to turn His vibrations into electrical signals and then back to vibrations. Since God also generates radio waves, the likelihood that His vibrations and light energy work together to produce meaningful language is significant. Humans have already figured out a similar process (live and recorded audio).

(6) Finally, the same biblical writer who stated that God is light also said God is love.[3] What if God's love has a physical manifestation—what if God generates love? Some scientists have theorized that one-dimensional strings may be what holds things together in our Universe. [Some scientists believe strings are the reality rather than particles, but

what if both strings and particles exist?] If there are strings that make up part of the reality of our Universe, they could add stability to all forms of matter. [Human love adds cohesion to our world in that it tempers competitiveness. Perhaps, God's love is the stability that turns chaos into order.] Also, some scientists have tried to develop the M-theory in their work surrounding the concept of strings. This hypothesis would explain the reality of everything, which could bring everything together in unity. What if this M-theory is the reality of God, the One who generates light, life, sound, and love? The Bible indicates that God is the reality of everything when the Apostle Paul wrote, "*And, we know that all things work together for good to them that love God ...*".[4] What God generates could be bundled together to work for good.

(7) As to where God resides: if He is a self-fueled, self-sufficient generator, He may not even dwell in the Universe. The reason why? God's light is so powerful that if He were present in the Universe, there would be no darkness. So, if God doesn't reside in the Universe, where does He dwell? Scientists have long flirted with the theory of parallel universes. Recently, scientists found evidence of light coming from what they think might be another universe. Could God's place of residence be what these scientists saw? The Bible indicates that God lives beyond what we see from here on Earth, as Paul described an experience he had had in 2 Corinthians 12. The apostle talked about being caught up unto the third heaven and named what he saw as Paradise.[5] That would work if we ascribe the Earth's atmosphere as the first heaven, the Universe as the second heaven, and God's dwelling place as the third heaven. Let's call that third heaven Infinity, where Eternity replaces time. These aren't new concepts for us—Infinity is boundlessness and endlessness, while Eternity is timelessness. [In the Bible, Isaiah referred to God as the "*high and lofty One who inhabiteth Eternity ...* ".[6]]

(8) We know that the Big Bang was when darkness came into existence and was the first step toward the development of the Universe. For darkness to appear, there had to be a light source first. What if that light source was God? What if the Big Bang happened within the place where God dwells, a place that we will call Infinity/Eternity (the antecedent to space/time)? The starting point for this new model is that God is a self-fueled, self-sustained generator of light, life, love, and vibrations who inhabits Infinity/Eternity. God's light, life, love, vibrations, and good filled Infinity/Eternity before the Big Bang.

Chapter Notes

[1] Revelation 4:5 KJV
[2] Psalm 36:9 KJV
[3] I John 4:8 KJV
[4] Romans 8:28 KJV
[5] 2 Corinthians 12:2-4 NASB
[6] Isaiah 57:15 KJV

Remember

None

Mini-Essay

Although there have been multitudes of scientific discoveries over the years, some scientists want to base reality on one specific theory. As of yet, they haven't been able to settle on which specific theory thoroughly explains everything. One particular difficulty lies with quantum mechanics and gravity. Scientists haven't been able to find a premise that bridges the gaps between how particles behave and how gravity exists. Maybe reality isn't one particular theory, but all of them combined. Let's think about God as a generator—what He produces would look like light. But, there's nothing simple about that light. God isn't one-dimensional, and what He generates isn't one-dimensional either. If we put everything He generates together, we will find multiple layers interacting. Perhaps, the scientists' quest to find the one "thing" that explains everything won't end with just one theory but with some of the theories combined. Perhaps, it isn't the generated products—energy, light, nuclei (particles), strings, vibrations—that are important in the theory of everything. Perhaps, the Generator of those products is the ultimate source of everything.

Since some scientists believe the string and loop theories might both be true, the two theories together could be a metaphor for the love relationship between God and an individual. If God manifests His love with one-dimensional strings, He could connect to each person via a string. Then, when that person decides to love God, the string loops back to God. The love relationship becomes a two-dimensional loop and creates a channel between the two who love. This channel opens a two-way communication which is the foundation of God's work in an individual's life. Without a choice to love God, what He generates remains general (examples: sunshine, rain, etc.) in a person's experience and not as personal "gifts" to them. So, it's logical that God wants us to love Him with everything we have so He can give us all that He generates. God isn't a selfish being who demands our loyalty; He is a mentor and helper who wants us to experience good. As individuals, we all have agency in loving God and creating the channel of love that will bring all He generates into our lives. Just a warning: if we say we love God while hating our fellow human beings, we become liars (1John 4:20 NASB). When we choose to love God, we also choose to love all human beings. Then our love relationships with God become multi-dimensional as we express our love for others through prayers and actions.

In thinking about God as a generator, a question arises: "Wouldn't He have to be humongous to generate enough to fill Infinity/Eternity?" Not necessarily so. In 2015, scientists discovered a ball of hot gas (far, far away from us), emitting energy equivalent to what would be emitted by hundreds of billions of our sun. These scientists determined that the object from which the hot gas is erupting is only ten miles across. The "generator" size has nothing to do with the amount of energy it produces. God may generate enough to fill all Infinity/Eternity and yet be relatively small and compact.

Suggested Reading

Cosmic Rays - Introduction. (2017, July).
https://imagine.gsfc.nasa.gov/science/toolbox/cosmic_rays1.html

Lamb, R., & Kiger, P. J. (2021, April 15). *What Existed Before the Big Bang.*
https://science.howstuffworks.com/dictionary/astronomy-terms/before-big-bang.htm

Magazine, Q. (2020, October 16). *String Theory Meets Loop Quantum Gravity.*
https://www.quantamagazine.org/string-theory-meets-loop-quantum-gravity-20160112/

Patel, N. V. (2015, October 29). *Scientists May Have Just Discovered a Parallel Universe Leaking Into Ours.*
https://www.inverse.com/article/7403-scientists-may-have-just-discovered-a-parallel-universe-leaking-into-ours

What is 10 miles across, but powers an explosion brighter than the Milky Way? (2016, January 14).

https://news.osu.edu/what-is-10-miles-across-but-powers-an-explosion-brighter-than-the-milky-way/

Wood, C. (2019, July 11). *What Is String Theory?* https://www.space.com/17594-string-theory.html

Zyga, L. (2019, July 2). *A counterintuitive case in which like charges attract.* https://phys.org/news/2019-07-counterintuitive

3

THE BIG BANG

Keywords: Evil, God, Bible, Light, Darkness, Tent, Universe, Funnel, Flared End, Drain End, Big Bang, Gluons, Quarks, Electrons, Photons, Phases, Empty, Reality, Agency

The Model

(1) Why does evil exist? This question has haunted us throughout our history. The answer goes back to Infinity/Eternity before the Big Bang; in fact, the need for evil was the catalyst for the Big Bang. Before God created any beings (entities that exist), He had to provide an alternative to Himself. Why? God is love,[1] and as 1 Corinthians 13:5 KJV states, love doesn't seek its own (way). As long as God was the only reality in Infinity/Eternity, any created being would have had to exist in His presence. Thus, God would be demanding His way without an alternative to Himself. This circumstance would have been contrary to the reality of who God is. So, before He created any other beings, God had to furnish an alternative to Himself. He had to engineer an area in Infinity/Eternity that would be absent of everything God is and generates. (We'll call that area the Universe.) That was the only way that there would be a legitimate choice for His created beings. Remember,[a] we talked earlier about how Isaiah quoted God saying, *"I form the light, and create darkness: I make peace, and create evil: I the Lord do all these*

things."?[2] This verse affirms God's decision to create an alternative to Himself. God is the generator of light, so He could create darkness by blocking Himself and everything He generates from the Universe. As far as God making peace and creating evil goes, conflict didn't exist in Infinity/Eternity because God was all-in-all. To provide an alternative to Himself, God had to create discord, which is the power of evil. All God generated was good (untainted, flawless), so He had to establish an opposing force. We can summarize the conflict: God (good) was constructive, while the alternative to God (evil) would be destructive.

(2) God faced two challenges in creating the Universe: cordoning off the Universe and emptying the space. First, God had to design a way to stop His generated cosmic rays from entering the Universe. Isaiah 40:22 NASB indicates what God chose to do: "*It is He ... who stretches out the heavens like a curtain And spreads them out like a tent to dwell in.*" So, let's think about God fabricating/erecting a tent-like structure that would enclose the Universe. He needed the structure to be conspicuous as a legitimate choice while assuring what He generated could not penetrate its boundaries. One possibility could have been a double-sided material with the outside coated with a substance that would absorb the vast majority of the light God generated. We know a coating with that amount of absorption is possible because human inventors have developed products from carbon nanotubes that absorb almost 100% of visible light. One of these paints, in particular, makes a three-dimensional object look like a black hole—if God used a similar substance in structuring the Universe, the area would look like a black hole from His perspective. Also, since the coating would have to have something to adhere to, God may have fashioned the other layer of the material as a strongly reflective surface. This underlying stratum would ensure that if anything seeped through the outer coating, it would reflect back through the coating and into Infinity/Eternity. So, God could have accomplished His purpose of cordoning off the Universe with the

double-layered material. [By the way, anything good that humans can create, so can God.]

(3) God's challenge of emptying the Universe of the things He generates centered on choosing the shape of the Universe. Over the years, scientists have debated that shape; one of their theories would be conducive to emptying the Universe of everything. Some scientists believe the Universe's configuration to be similar to an old-style trumpet with a long stem and a flared end. [Perhaps we can modify this theory by describing the shape as a funnel with a flared bell and a narrower drain end.] As scientists worked on this theory, one discovery perplexed them. They considered the trumpet shaped-universe in terms of people traveling to the edge of the flared bell. The scientists figured out that, instead of leaving the Universe, the travelers would find themselves turned around and going in the opposite direction (back into the Universe). The answer to that enigma could be that God sealed the flared end with a lid. Let's think about that—God had to stop the flow of His energy, light, life, and sound (in the form of cosmic rays) from entering the Universe to assure the integrity of the alternative to Himself. The presence of a lid would be part of what God worked out and would explain this phenomenon that has been puzzling scientists. As for the drain end, God would have then had to seal it to assure that nothing He generated would seep back into the space. He also had to prepare for the scenario that some of the first created beings in Infinity/Eternity would choose the alternative He had created. In that scenario, the drain end would become unsealed, which would open the door to everything God generates seeping back into the space. [Since the alternative to Himself would already be in place, God could begin a new creation within the Universe and still be true to His character.] The reflective surface of the inner layer of the boundary would facilitate the effects of everything that leaked back into the area.

(4) When God capped and sealed the flared end of the funnel, He stopped the flow of His energy and everything He generates.

Remember,[b] scientists have found that particles with like, but unequal charges can attract each other. As the flared end of the funnel was covered, the energy and the charges within the space began to weaken —the flow from God completely stopped. As the energy level waned, the particles with the weakened charges moved through the drain end towards the fully-charged particles in Infinity/Eternity. In particular, the gluons holding the nuclei's protons and neutrons together lost their strength. When that happened, the protons and neutrons fell apart into quarks which became the residue of the Big Bang. To understand how that might have happened, let's consider the possibility that photons, electrons, and gluons are phases of the same particle, much like steam, water, and ice are phases of H_2O. The three stages of H_2O have differing physical qualities, but they all "flow" from one state to another under certain circumstances. If the same is true for photons, electrons, and gluons, as the Big Bang happened, the gluons holding the quarks together may have transformed into electrons and drained out of the Universe. Another important point: because the flared end of the funnel has a much greater volume than the drain end, the onslaught of the particles flowing out of the Universe could have overwhelmed the drain end if the substance lining the drain had been just the reflective layer. If God lined the inner drain area with the same absorbent material as the outer layer, He could ensure the mechanics of draining the funnel would work smoothly, and the Universe could be free from all that God generated.

(5) So, after everything drained out of the Universe, God sealed that end of the funnel. The space was dark, quiet (no vibrations), cold, and empty of everything but quarks (the ashes left behind). Here's where this model diverges from what scientists have theorized about the after effects of the Big Bang. Some of them believe the new Universe was a scorching hot and dense combination of gluons and quarks. The scientists believe that there had to be energy present for quarks and gluons to recombine into nuclei, for electrons and nuclei to form into atoms, and for the atoms to form into molecules. By not understanding

what existed before the Big Bang, scientists can't theorize anything other than the hot and dense mixture being its result. However, when we comprehend who God is and His purpose in creating the Universe, the absence of everything but quarks makes sense. A cold, empty Universe was the exact opposite of Infinity/Eternity and was thus a viable alternative to God.

Chapter Notes

[1] 1 John 4:8 NASB
[2] Isaiah 45:7 KJV

Remember

[a] Chapter 2, Paragraph 3
[b] Chapter 2, Paragraph 2

Mini-Essay

Isaiah 45:7 raises an essential question about God, "Is He a wishy-washy tyrant, a two-headed monster?" Why would the good God create evil? Let's put 1 Corinthians 13:5 (love doesn't demand its way) together with Deuteronomy 30:15-20, both KJV. We can infer that evil exists because God isn't a puppet master, so He wanted to make sure we had a legitimate choice in what was available to us. Let's look at most of the Deuteronomy passage:

> *See, I have set before thee this day life and good, and death and evil; in that I command thee this day to love the Lord thy God, to walk in his ways, and to keep his commandments and his statutes and his*

> *judgments, that thou mayest live and multiply: and the Lord thy God shall bless thee in the land whither thou goest to possess it.*
>
> *But if thine heart turn away, so that you will not hear, but shall be drawn away, and worship other gods and serve them; I denounce unto you this day, that you shall surely perish, and that ye shall not prolong your days upon the land, whither thou passest over Jordan to go to possess it.*
>
> *I call heaven and earth to record this day against you, that I have set before you life and death, blessing and cursing: therefore choose life, that both thou and thy seed may live: that thou mayest love the Lord thy God, and that thou mayest cleave unto him: for he is thy life, and the length of thy days ...*

Even as early as the time of Moses, God revealed the reality of the choice we individually have.

We have agency with God. Because He doesn't demand His way, we are free to choose whether to love Him or not. As the Deuteronomy passage lays out, our choice is stark: life or death, blessings or cursings. Our choices (not God's) determine our experiences. The biblical writers mostly did not understand this concept. They equated destruction with God's Wrath (God being angry), so if something horrible happened, God was responsible for it. They believed God punished people when they disobeyed Him. That's not it. God doesn't change what He does according to what we do. James 1:17 NASB says: "*Every good thing given and every perfect gift is from above, coming down from the Father of lights, with whom there is no variation or shifting shadow.*" God always generates energy, light, life, love, vibrations, and good. All of these work to our good and build us up into life. We often don't experience the good because we choose the alternative to God. He wants us to cleave to Him because He wants us to benefit from everything He generates.

Suggested Reading

Janek, V. (2015, May 12). *What shape is the Universe?* https://phys.org/news/2015-05-universe.html

Kooser, A. (2019, September 13). *Vantablack swallows everything it's painted on.* https://www.cnet.com/pictures/vantablack-swallows-everything-its-painted-on/

Zyga, L. (2019, July 2). *A counterintuitive case in which like charges attract.* https://phys.org/news/2019-07-counterintuitive-case.html

4

THE HOUSE OF MIRRORS

Keywords: God, Universe, Infinity/Eternity, Angels/Stars, Lucifer, Migrants, House of Mirrors, Cosmic Rays, Fuel, Quarks, Photons, Electrons, Gluons, Nuclei, Matter, Antimatter, Dark Matter, Reflection, Time, System

The Model

(1) After God sealed the Universe, Infinity/Eternity had two distinct areas: the place where God resided and the place that was utterly free of God and everything He generates. Since a definite choice existed, God was free to create the first beings, which the biblical writers named angels. In Psalm 104:4 KJV, the writer described the essence of these first beings, "*Who (God) maketh his angels spirits, his ministers a flaming fire.*" This verse characterizes the angels as being able to generate energy (spirits) and light (flaming fire), which would establish the angels as star-like beings. Later on in the Bible, Isaiah talked of the angels as the stars of God while he described Lucifer and his actions, which included vibrations (speech).[1] We can infer from these passages that the first created beings generated energy, light, and vibrations like God generated. However, there was one huge difference between God and

His angels/stars. God relies on no other being, force, or fuel to exist; He is self-sufficient. We can infer the angels/stars were not self-fueled or self-sufficient because, if they had been, they would have existed apart from God. Also, let's remember[a] that within the cosmic rays God generated were the nuclei of the elements, including hydrogen (the most abundant element in cosmic rays). As scientists have discovered, stars in the Universe burn hydrogen. We could use that fact to hypothesize that the angels/stars would cease to exist unless they received hydrogen from God through the cosmic rays that fill Infinity/Eternity. They would become nothingness. God was not only their creator, but He also was their sustainer.

(2) Now, let's think about Lucifer. Using different figurative language than Isaiah, the prophet Ezekiel referred to Lucifer as the created and anointed cherub who walked back and forth amid fiery stones (the other angels/stars).[2] From Ezekiel's reference, we can infer that the angels/stars had freedom of movement in Infinity/Eternity and were not bound to remain in specific spots. This ability is essential to understand as we look at the events transpiring after the creation of the angels/stars. Once He created the angels/stars, God gave each of them a choice between Himself and the Universe (like He gave Adam and Eve a choice between life and death[3]). God did not restrain Himself in giving His newly created beings understanding and wisdom—Ezekiel states that Lucifer was full of wisdom but that the angel corrupted his insight because of his brightness.[4] As Isaiah related, this corruption happened because Lucifer wanted to be like God and have a throne (position) higher than the other stars.[5] He could never attain those goals as long as he resided where God was, so he looked to the Universe. Even though God warned Lucifer that whoever migrated to the Universe would become nothingness, Lucifer was willing to let go of God and everything God generated to achieve his goals. Not only was he ready to become the exact opposite of God, but he had also organized a sizable group of angels/stars to go with him.[6] Lucifer wanted to be like God.

(3) So, Lucifer and his followers decided to migrate to the Universe, which was cold and empty. With the side of the boundary facing the Universe being reflective, their destination could have been considered a house of mirrors. Adding the perpetuity of Infinity/Eternity (including the Universe) to the mix made the Universe immeasurable for two main reasons: 1.) the inability to tell what was the reality and what was a reflection and 2.) the lack of a way to compute existence. Thus, when Lucifer and the other angels/stars entered the Universe, each arrival became a point or event in Eternity and was "recorded" by the reflective surfaces of the Universe's boundary. (We could compare our view of the Universe to a diorama rather than to a video recording.) The angels/stars had to enter through the drain end to keep the cosmic rays from flooding back into the space. Lucifer assumed the absorbent lining of the drain would take care of any seepage into the Universe. Also, Lucifer knew that God would never enter the enclosed area Himself because that would force any beings living in the Universe to choose God, and God would never do that. How could Lucifer have known these things? God is consistent—He never changes.[6] God never stops generating vibrations that communicate reality. It stands to reason that God warned the angels/stars of the consequences of choosing the Universe (consequences which included God's intent to create more beings within the Universe if Lucifer breached its boundary). However, we can infer that Lucifer was not intimidated; he was so caught up in his beauty that he was confident he could bring any newly created beings under his authority, as long as God didn't enter the Universe. Lucifer believed that those beings would worship him because he would be the brightest being in that space.

(4) As Lucifer and his followers entered the drain end of the Universe, the first effect was the absorption of the light they were generating. Since there were no cosmic rays inside the space, their fuel began to run out immediately. Scientists have documented how stars die, partly because the reflective boundary of the Universe has recorded each step of the way. As these angels/stars died, they degenerated into

invisible blobs of nothingness, what scientists call black dwarf stars. Some angels/stars died with massive explosions, while others faded away. The reflective surfaces captured and multiplied images of the migrants' transformations into the blobs of nothingness as they lost their existences as God's angels/stars. After they all filed in, Lucifer couldn't reseal the drain end of the space, and God didn't intervene because they had made their choice. He could begin to create within the boundaries of the Universe since the occupied space provided a legitimate choice between God's Somethingness and Lucifer's nothingness. So, everything God generated began seeping back within the boundaries of the Universe. If we look at Genesis 1:1-3 KJV, we can see the process of creating as it originated: "*In the beginning God created the heaven and the earth. And the earth was without form, and void; and darkness was upon the face of the deep. And the spirit of God moved upon the face of the waters. And God said, Let there be light: and there was light.*" [The spirit of God equals energy; the words God spoke, vibrations; and the light God generates, frequency.] Everything needed to recombine the quarks into protons and neutrons, including the flow of electrons, was seeping into the space. In a sense, the funnel of the Universe became a capped megaphone, where what filtered in was amplified through reflection. And so, the reversal of the Big Bang began.

(5) With the influx of energy, light, and vibrations into the Universe, matter began forming as the quarks recombined into protons and neutrons, which happened as some of the incoming photons and electrons started phasing into gluons. But, this recombination was not the only way matter came into being in the Universe. The cosmic rays that carried all that God generated were seeping into the space. [As we've discussed previously, cosmic rays carry the nuclei of all known elements.[b]] Both sets of nuclei—the recombined and the new—swarmed in the mixture of energy, frequencies, and vibrations rich with electrons. Atoms and then molecules came together, and the first matter in the Universe was "created." At this point, it's important we

understand that scientists have theorized that three "aspects" of matter exist: matter itself, antimatter, and dark matter. They define antimatter as the mirror image of regular matter. The theory is that when the particle of matter collides with its corresponding antimatter, the two particles obliterate each other and release a gamma wave. We need to be careful here because antimatter may be an illusion rather than a reality. If the Universe is rich with reflective surfaces, the mirror images of different particles may be reflections of the particles. When they hit a reflective surface, both electrons and gluons may phase back to the photon state. Since the only way that scientists "detect" antimatter is the production of gamma rays (energetic photons), they may be seeing particles hitting the reflective surface, thus causing electrons and gluons to revert to photons. So, if photons, electrons, and gluons are the same particles in different phases, antimatter may not exist. Also, as far as matter and dark matter go, let's rename them Somethingness (God) and nothingness (Lucifer).

Chapter Notes

[1] Isaiah 14:12-13 KJV
[2] Ezekiel 28:14, KJV
[3] Genesis 2:16-17 KJV
[4] Ezekiel 28:12, 17 KJV (Lucifer is alluded to as the King of Tyre.)
[5] Isaiah 14:13-14 KJV
[6] James 1:17 KJV

Remember

[a] Chapter 2, Paragraph 4
[b] Chapter 2, Paragraph 4

Mini-Essay

Before we go on, let's think about the concept of time. Some people take Genesis 1 literally and believe that God created everything in six twenty-four-hour days. How ironic is it that we who should be aware of Eternity are not? When thinking about the Creation, we tend to forget that God is eternal, that He isn't bound by time. We also forget that God had a specific reason for setting up a time structure. Genesis 1:14 KJV states that the sun and moon—the foundation of time—were put in place for signs, seasons, days, and years. In other words, time exists so human beings can keep track of things. Time doesn't exist for God.

Additionally, those who believe in the six-day Creation push God into the role of hypocrite and set themselves up for failure. Think about it. How many times in the Bible are we admonished to be patient and wait for God to work? If He only took six days to create an incredibly complex and intricate universe (complete with biological life), where does He get off telling us to be patient? If it only took Him six days to achieve His goals, why can't He answer our prayers immediately? By not understanding the eternal aspect of Creation, our expectations of God are skewed. When we don't see our prayers answered promptly, our faith can take a battering. We need to dispense with the concept that God created within a time frame. To go one step farther, Creation hasn't stopped because God still generates, and matter is still forming as the cosmic rays enter the Universe. [The biblical writers had no comprehension of the Universe, so they could only write about the Creation in terms they understood.]

Scientists don't get off scot-free in dealing with the concept of time. They take a system that has been put in place for humans to keep track of things here on the Earth and try to transpose it to help measure what is immeasurable. We can't measure Eternity in years; when we talk about billions of years, we're talking Eternity. And, since movement—

the Earth on its axis and the Earth's orbit around the sun—determines time, we need to consider when that movement started. Most things made/manufactured for movement (vehicles, etc.) don't move until they come off the production line. What if time didn't begin until after the basic Creation process described in Genesis 1 ended? Can we measure the age of the Earth with a system that didn't exist while the Earth was forming?

We also need to look at another concept scientists have adhered to: the Universe is a closed system. This foundational understanding may be crumbling with the discoveries 1.) of more light in the Universe than can be accounted for by known galaxies and 2.) of light that may be coming from a different "universe." This new information indicates an inflow into the Universe which would make it an open system.

Suggested Reading

New Horizons Spacecraft Answers Question: How Dark Is Space? (2021, January 12).
https://hubblesite.org/contents/news-releases/2021/news-2021-001

Patel, N. V. (2015b, October 29). *Scientists May Have Just Discovered a Parallel Universe Leaking Into Ours.*
https://www.inverse.com/article/7403-scientists-may-have-just-discovered-a-parallel-universe-leaking-into-ours

The matter-antimatter asymmetry problem. (2021, April 14).
https://home.cern/science/physics/matter-antimatter-asymmetry-problem

5

BLACK HOLES, WORMHOLES, EXPANSION, CIRCLES, GRAVITY, AND A DICHOTOMY

Key Words: God, Somethingness, Universe, Lucifer, Nothingness, Black Holes, Gravity, Wormholes, Vibrations, Infinity/Eternity, Slithering, Dark Energy, Expansion, Reflective Surface, Cosmological Constant, Acceleration, Circles, Curvature, Strings, CMBR, Magnetic Fields, Iron, Radiation Friction, Dichotomy

The Model

(1) As God's Somethingness seeped into the Universe, the resulting matter started exerting pressure on Lucifer and his followers. They didn't change in any way—they were still blobs of nothingness—but their environment was undergoing a tremendous adjustment. As matter formed, it pushed the blobs closer together, and some started melding into what scientists now call black holes. Scientists believe

black holes are places in space where gravity pulls so much that light cannot get out. But, what if black holes have nothing to do with gravity, and what if there are no frequencies (light) inside them? [Remember,[a] the invisible blobs of nothingness came into being as Lucifer and his followers entered the Universe and no longer generated light.] Scientists believe that black holes are super dense, which could negate this concept that the black holes are full of nothingness. However, the Bible may explain how these holes can be dense and full of nothingness at the same time. In Job 40, God told Job about the leviathan, a metaphor for Lucifer (aka Satan), who destroyed Job's life. Let's look at part of what God said in Job 40:15-17 KJV:

> *"His (the leviathan's) scales are his pride, shut up together as with a close seal. One is so near to another that no air can come between them. They are joined one to another, they stick together, that they cannot be sundered."*

Applying this metaphor of scales to Lucifer's followers, they could be packed so tightly together in the black holes that matter can't penetrate them nor can light travel through them. And, since scientists believe black holes grow in size, perhaps the still-developing matter in the Universe continues to exert force on the blobs of nothingness.

(2) Along with the concept of black holes, scientists theorize that wormholes also exist and may connect dark holes. Such a set-up makes sense because Lucifer needed to communicate information within his kingdom through vibrations while he and his followers also needed to travel throughout the Universe. In thinking about how the blobs of nothingness move, the Bible gives us a clue. In Genesis 3:14, God inferred the only type of movement Lucifer and his followers had was the movement of serpents (snakes, worms, etc.). As they migrated from Infinity/Eternity to the Universe, these angels/stars went from the soaring as stars to the slithering as worms. Thus, it's reasonable to

hypothesize that the blobs created a network of wormholes. It's also logical to think that the vibrations and slithering movements of Lucifer and his followers could be what constitutes the dark energy that scientists believe exists in the Universe. The scientists connect dark energy with the expansion of the Universe that they think is happening.

(3) Let's deal with the theory of the Universe expanding—that may not be a reality at all. Because the boundary between the Universe and Infinity/Eternity may consist of a reflective surface, our telescopes could be picking up an infinite number of images of the same events (think of infinity mirrors). The reflections could be why the Universe seems to be rapidly expanding. The Universe could be a stable, cordoned-off area in Infinity/Eternity and could be neither expanding nor contracting—it could occupy an unchangeable, finite space. We could say the Universe's size is constant. And concerning the theorized dark energy, scientists have been looking for a cosmological constant to reconcile what they theorize with what they observe in the Universe. They see the constant of being a number representing a force that is anti-gravity (thus causing the Universe to accelerate its expansion), and they label that force as dark energy. However, the dark energy may be the movement of the blobs of nothingness within the wormholes that causes matter to move about with a continued acceleration. And, the faster matter bumps up against the black holes, the more violent the turbulence at the event horizons becomes.

(4) With its funnel shape, the Universe is circular, and the objects and patterns inside it are elliptical/circular. If we could peer into the Universe from the outside, we would see multitudes of ellipses and circles that rotate and then revolve around each other. Curvature seems to be the default setting of the Universe and of much of what occupies the space. Even black holes and wormholes, though unseen, are thought to be elliptical/circular. All the spinning ellipses/circles of matter could be the foundation of a system that produces gravity. Other components of that system could include the (theorized) strings,

the Cosmic Microwave Background Radiation (CMBR) that permeates the Universe, the magnetic fields, and the abundant presence of iron throughout the space.

(5) First, strings could give both stability and flexibility (think of the rope attaching a tetherball to a pole), which would allow one object to revolve around another object. Then, the CMBR could provide radiation friction that could generate the magnetic fields (that attract) to offset the dark energy (that pulls apart). Finally, iron is easily magnetized—as it is abundant throughout the Universe, it could make the magnetic fields work. Also, we could infer that both the CMBR and dark energy permeate the Universe and simultaneously exist in the same space. We could say that light and darkness fill the Universe, yet they are separate entities. What a dichotomy!

Chapter Notes

None

Remember

[a] Chapter 3, Paragraph 4

Mini-Essay

Let's meld together the concepts of strings and empty black holes/wormholes. If the theorized strings are the reality, then matter could be considered the fabric of the strings woven together. We could view the empty areas as holes in the fabric. The dark energy of the slithering blobs of nothingness bumping into matter causes the fabric to ripple.

Also, if we gathered the fabric up and spread it out, we would see the holes where matter doesn't exist. [Here's a paradox: even though the black holes are free from matter, the blobs of nothingness within them do exist and vibrate.] Since some scientists believe that consciousness comes with vibrations and that every iota of matter vibrates, we can infer that all matter has consciousness. With two distinct sets of vibrations in the Universe, we can surmise that two different consciousnesses—vibrations both from God and the blobs of nothingness—exist within its boundaries. Let's think about this as we work through the creation of life and then pick the topic up again when we discuss the Garden of Eden.

Christ once said, "*Truly, truly, I say unto you, before Abraham was born, I am ...*" and caused the religious people to want to stone him (John 8:58-59 NASB). The Jews limited the name "I Am" to God and thought it blasphemy to connect that name to a human being. Because they had no concepts of the physical realities of the Universe (or even the reality of the Universe itself), they saw God as the only eternal existence. What Christ said about himself is true of all of us and every iota of matter. What God generates is eternal and will never cease. So, before the "parts" came together, we already were. Because God exists, we exist. Because God is, we are.

Suggested Reading

Dark Energy, Dark Matter. (2021, June 22). https://science.nasa.gov/astrophysics/focus-areas/what-is-dark-energy

Hunt, T. (2021, October 14). "'*Resonance Theory'—Could Consciousness all Come Down to the Way Things Vibrate?*"

https://scitechdaily.com/resonance-theory-could-consciousness-all-come-down-to-the-way-things-vibrate/

Johnston, H. (2017a, August 29). *"Radiation friction" could make huge magnetic fields with lasers –.* https://physicsworld.com/a/radiation-friction-could-make-huge-magnetic-fields-with-lasers

Redd, N. T. (2017, October 21). *What Is Wormhole Theory?* https://www.space.com/20881-wormholes.html

Wikipedia contributors. (2022, February 14). *Cosmological constant.* Wikipedia. https://en.wikipedia.org/wiki/Cosmological_constant

Wild, F. (2018, August 22). *What Is a Black Hole?* https://www.nasa.gov/audience/forstudents/k-4/stories/nasa-knows/what-is-a-black-hole-k4.html

6

IT'S ALL ABOUT THE DNA

Key Words: DNA, Natural Selection, Common Ancestor, Diversity, God, Amoeboid Family, Clone, Organism, Chromosomes, X-chromosome, Y-chromosome, Reproduction, Genome, Base Pairs, Mitochondrial Eve, Y-chromosome Adam, Gametes, Siblings, Individual, God's Love, Judgment, Unloving, Unmerciful

The Model:

(1) To explain the diversity of life forms and the evidence of the extinction of other life forms, scientists developed the theory of natural selection in the nineteenth century. They hypothesized that all organisms share a common ancestor and that the different organisms evolved by chance, with the ones with the hardiest traits surviving to reproduce. Since then, scientists have built on this basic understanding, but they've missed some essential questions to ask to understand the diversity of life:

- How did DNA originate?
- How did the nuclei (carbon, hydrogen, oxygen, nitrogen, sulfur, and phosphorus) become atoms? How did those atoms configure themselves into nucleotides and DNA molecules?
- How did the nucleotides know how to pair up?
- How did the histones know how to wrap genes into chromosomes?

And the questions go on and on. As any creative person can testify, the medium and materials of a new product/project do not randomly meld together into the new creation. At least two "forces"—directed energy and the intentional manipulation of the medium and materials —are needed to bring any new work into being. Without these two dynamics, the raw materials would remain separated, no matter how long they were available to be used. Likewise, nucleotides and histones would never meld together from the six elements and then arrange themselves into DNA without directed energy and intentional manipulation. Somehow, there was a source of the focused energy and the deliberate manipulation needed to build the building blocks of life.

(2) The theory of natural selection has been the foundation that undergirds much modern scientific thinking. But, in light of the need for directed energy and intentional manipulation to create new things, this concept of massive randomness seems similar to the magical thinking of God creating the Universe in six twenty-four-hour days. That said, the Bible does explain two ways that God created life on Earth. Here is what Genesis 2:7, 19, 21-22 KJV says:

> *And the Lord God formed man out of the dust of the ground, and breathed into his nostrils the breath of life; and the man became a living soul … And out of the ground the Lord God formed every beast of the field, and every fowl of the air … And the Lord God caused a deep sleep to fall upon Adam and he slept: and he took one of his ribs, and*

closed up the flesh instead thereof; and the rib, which the Lord God had taken from man, made he a woman, and brought her unto the man.

[Although these descriptions centered on animal life, we can infer the creation process for plant life forms was similar.] These verses show the methods God used to build biological life on Earth: creating a new form and adapting an existing form by manipulating its DNA (the rib bone). Before we look at the process God may have used, let's consider a few scientific findings:

- In 2001, scientists estimated that one species of the single-cell amoeboid family had 670 billion base pairs of DNA in its genome while another species had 290 billion base pairs (a human has about 3 billion base pairs). Although the numbers may be too high due to outdated measurement methods, these amoeboids have gigantic genomes for their size.
- Some amoeboids clone themselves, which means they are "immortal" unless killed or starved. Their genomes pass on to their daughters for many generations.
- Some scientists believe all biological life forms have the same single one-cell ancestor.

So, let's see how the Bible passage and the scientific observations work together.

(3) In one possible scenario, God created enough DNA (whether on the Earth, in the Universe, or in Infinity/Eternity) to make a diverse world; He also created the cell structure to house the DNA. When He put the DNA in the cell, a living organism came into being and stared cloning itself. To develop a variety of life on the Earth, God used individual clones of the original cell. For each type of new organism, He took one clone, culled the DNA in the nucleus (think laser surgery), and put the single cell in a mud "womb," where it began to divide and

develop into a new life form. And, since God needed each life form to have the capability to reproduce, He set up the male/female system (with some exceptions), which usually involved two different chromosomes. For each new life form, the original cell contained both sex chromosomes. After it had fully developed into a mature life form, God took the DNA of the life form and made some adjustments, especially to the sex chromosomes. For example, the male human has both the X and Y chromosomes, so God "created" a man first. Then, He took some of the man's DNA, removed the Y chromosome, and duplicated the X chromosome. Voila, there was the opposite sex and the capability of reproduction. [The females have both sex chromosomes in some life forms, so they would have been the original life form. In still other life forms, one sex has two identical chromosomes while the other has one, meaning the original organism could have been either sex.]

(4) In light of the burgeoning understanding that scientists are gaining about DNA, a particular verse in the Bible becomes more relevant. Scientists can write human genomes out in letters, with the nitrogenous bases coming in at over six billion letters (over three billion base pairs) alone. The letters of the base pairs in one human genome, written in the order as they appear on the DNA strand, would fill almost 300,000 printed letter-sized pages. This reality gives new meaning to Psalm 139:16 NASB: "*Your eyes have seen my uninformed substance; and in Your book were all written the days that were ordained for me, when as yet there was not one of them.*" God sees human DNA and knows who people are before they have developed in their mothers' wombs. Using this concept, we can infer God's Book of Life contains the genomes of all living organisms that have been or are on Earth. [Without the DNA, there would be no life on our planet.] From one perspective, we could say the life forms on Earth are the Words of God because they originated from Him, and we can write their genomes out. And, except for offspring created through asexual reproduction (aka clones), each individual is unique in its DNA because it receives its genetic

material from both parents. God's Book of Life could be close to being infinitely long.

(5) Scientists have concluded that all living humans are descendants of one "mitochondrial Eve," and all living men are descendants of one "Y-chromosome Adam." For this model, let's infer that the first two human beings were the mitochondrial Eve and the Y-chromosome Adam. Neither of those humans had parents, so their DNA had no lineage. Thus, Eve and Adam's children were not related to each other like today's siblings are. When we look at the diversity we see in the human race, the differences among people may have been inscribed in the genes carried by the gametes of both Eve and Adam. In other words, each gamete may have had a slightly different genetic code in its single-strand DNA with enough variants to create the vast diversity we see. This same principle applies to all life forms that came into existence.

(6) Since gender is such a massive issue in today's world, let's look at some of the observations scientists have made in their research:

- Many variations to the XX and XY gene pairings (that somewhat determine gender) exist, with some of the combinations going beyond just two primary chromosomes. Included are such combinations of the XX male, XY female, XXY, XXX, XYY, XXYY, etc.
- A high percentage of the genes involved in the differences between males and females don't appear on the X and Y chromosomes.
- When an individual carries two or more X chromosomes, all but one are inactivated and reside in a Barr body. However, some of those genes escape the Barr body and can become partially expressed in cells, contributing to sex differences in the individuals.
- About 99% of DNA is noncoding—scientists used to think it was junk, but now they believe that at least portions of it are integral to the function of cells.

- The human body has trillions of chemical reactions going on simultaneously. The individual's DNA regulates these reactions and directly expresses their genetic makeup (including sexual identity).

How our DNA makes each of us an individual with a unique gender goes way beyond a simple combination of two of the sex genes. Scientists have barely scratched the surface of fully understanding DNA and how it governs life forms. There is so much more to comprehend about the intricate complexity that composes biological life here on the Earth.

Chapter Notes

None

Remember

None

Mini Essay

As we learn how each one of us is a unique individual, perhaps Galatians 3:28 KJV can be looked at from a new perspective: "*There is neither Jew nor Greek, there is neither bond nor free, there is neither male nor female: for you are all one in Christ Jesus.*" God doesn't see us as groups—He sees and loves each of us as a unique individual. We are all one in Christ because Christ explicitly expressed God's love for all of us. [Another dichotomy: We are individuals, and we are one through the love of God.] Perhaps, we should take greater care in how we treat ourselves and our fellow human beings. We might want to avoid negating God's love for everyone by pronouncing judgments upon other people.

The Apostle Paul warned against judging others in the first two chapters of the book of Romans. Some Christians pick out Romans 1:26-27 to condemn homosexuality and stop there. We need to keep reading because Paul adds the unloving and unmerciful to the list of people who have depraved minds. Unloving means demanding our way—it's trying to force others to live the way we think is right. Unmerciful means being cruel or harsh—it's condemning others who don't live the way we think is right. We would do well to read Romans 2:1-4 NASB:

> *Therefore you have no excuse, everyone of you who passes judgment, for in that which you judge another, you condemn yourself; for you who judge practice the same things. And we know that the judgment of God rightly falls upon those who practice such things. But do you suppose this, O man, when you pass judgment on those who practice such things and do the same yourself that you will escape the judgment of God? Or do you think light of the riches of the kindness and tolerance and patience, not knowing that the kindness of God leads you to repentance?*

Paul is saying that the very act of judging another person (religious judgments, not legal judgments) is the same as doing what we're condemning. By judging others, we put ourselves in the arena of being unloving and unmerciful, which equals a depraved mind. Remember, Christ told us to judge ourselves rather than others (Matthew 7:1-5 NASB).

Suggested Reading (and Watching)

IMPORTANT VIDEO: *DNA Doesn't Look Like What You Think!* (2017, September 5).
https://www.youtube.com/watch?v=aT9T9h-2Jp8

46,XX testicular disorder of sex development: MedlinePlus Genetics. (2020).
https://medlineplus.gov/genetics/condition/46xx-testicular-disorder-of-sex-development/#causes

Dance, A. (2020, March 20). *Genes that Escape Silencing on the Second X Chromosome May Drive Disease.*
https://www.the-scientist.com/features/genes-that-escape-silencing-on-the-second-x-chromosome-may-drive-disease-67124

Frith, R. (2021, June 16). *Variations.* The Association for X and Y Chromosome Variations.
https://genetic.org/variations/

Genetic Mechanisms of Sex Determination | Learn Science at Scitable. (2014).
https://www.nature.com/scitable/topicpage/genetic-mechanisms-of-sex-determination-314/

Ghose, T. (2013, August 1). *Genetic "Adam" and "Eve" Uncovered.*
https://www.livescience.com/38613-genetic-adam-and-eve-uncovered.html

How Many Pages Does It Take to Print Out Someone's Genetic Code? (Video). (2016, May 6).
https://www.kqed.org/futureofyou/154916/how-many-pages-does-it-take-to-print-someones-genetic-code-video

NHGRI. (2019, March 9). *Deoxyribonucleic Acid (DNA) Fact Sheet.* https://www.genome.gov/about-genomics/fact-sheets/Deoxyribonucleic-Acid-Fact-Sheet

Nugteren, A. (2018, May 13). *One Second in Your Body.* https://scientiapotentiaest.blog/2018/05/14/one-second-in-your-body/

Perman, A. (2014, December 4). *The world's largest sequenced genome is just the start.* (Check out the infographic.) https://blogs.biomedcentral.com/on-biology/2014/03/20/worlds-largest-sequenced-genome-unlocking-the-loblolly-pine/

Staff, S. X. (2018, May 21). *Far from special: Humanity's tiny DNA differences are "average" in animal kingdom.* https://phys.org/news/2018-05-special-humanity-tiny-dna-differences.html

Than, K. (2021, May 4). *All Species Evolved From Single Cell, Study Finds.* https://www.nationalgeographic.com/adventure/article/100513-science-evolution-darwin-single-ancestor

Than, K. (2021, May 17). *What is Darwin's Theory of Evolution?* https://www.livescience.com/474-controversy-evolution-works.html

What is noncoding DNA?: MedlinePlus Genetics. (2021, January 19). https://medlineplus.gov/genetics/understanding/basics/noncodingdna/

Wikipedia contributors. (2021, December 26). *Sex-determination system.* https://en.wikipedia.org/wiki/Sex-determination_system

7

THE GREAT SCAM

Key Words: Garden of Eden, Vibrations, Sound Waves, Resonance, Consciousness, Constructive Interference, Destructive Interference, God, Nothingness, Synchronized, Matter, Amplitude, Somethingness, Lucifer, Nothingness, Adam, Eve, Fruit, Knowledge, Good, Evil, Deception, Commandments, Lack, Experience, Opposite, Victims, Religion, Consequences, Dark Energy, Death, Destruction, Loss, Sin, Choice

The Model

(1) Before we can understand what happened in the Garden of Eden, we need to visit the scientific concepts of vibrations, sound waves, resonance (the resonance theory of consciousness), constructive interference, and destructive interference:

- Remember,[a] both God and the blobs of nothingness vibrate and thus produce sound waves.
- When sound waves are synchronized (when they coincide in timing and rate), they produce resonance.
- Some scientists believe consciousness and resonance are intertwined and present in all matter, even the smallest particles.

- Sometimes sound waves from different sources can be at the same place simultaneously. If the waves are in-phase, constructive interference happens as the waves amplify each other.
- If the waves are out-of-phase and have exact opposite amplitudes, destructive interference produces a wave with zero amplitude.

All of these concepts impacted what happened in the Garden of Eden.

(2) God is consciousness—He exists; therefore, He experiences. (One can't experience unless one exists, and one can't exist without experiencing.) Since matter exists because of what God generates, it makes sense that matter also has consciousness, the consciousness of Somethingness. Meanwhile, Lucifer and his followers also have consciousness, but their consciousness is the exact opposite of God's consciousness. [As Lucifer and his followers migrated to the Universe, they lost all consciousness of God and took on the consciousness of nothingness.] After God established Adam and Eve in the Garden of Eden, their only consciousness was of His Somethingness; they only knew good at that point and had no concept of anything else. However, God warned the humans of the presence of the second consciousness (evil) and the consequences of eating the fruit of nothingness[1]: if they ate the fruit of the tree of the knowledge of good and evil, death (the result of evil) would follow. Until Lucifer approached Eve in the guise of a serpent, the humans didn't touch the fruit of nothingness. When Lucifer spoke to Eve, his words were the exact opposite of God's words. [Keep in mind the words Christ spoke about Lucifer in John 8:44 NASB: *"You are of your father the devil (Lucifer) ... Whenever he speaks a lie, he speaks from his own nature, for he is a liar and the father of lies."*] Let's look at the gist of Lucifer's interaction with Eve as described in Genesis 3:1-7 NASB:

- Lucifer's first words were a deception disguised as a question: *"Indeed, has God said, 'You shall not eat from any tree of the garden'?"*

God had told Adam and Eve to eat from all but one tree.[1] After Eve told Lucifer about the tree of the knowledge of good and evil, the first thing he responded with was, *"You surely will not die."* This statement is a direct contradiction to what God had told Adam.

- Then, Lucifer continued with the words, *"For God knows that in the day you eat from it your eyes will be opened ... ."* The humans had already seen the goodness of God. The only thing that they hadn't seen was the evilness of nothingness. That nothingness would lead them to be blind to God's goodness. Eating the fruit would cause blindness. Lucifer lied again.
- Next came the clincher: *"... and you will be like God ..."* Lucifer's greatest desire was to be like God and rule the Universe. Since God had already given Adam and Eve dominion over the animal kingdom, Lucifer could not use that perk to convince the humans to eat the fruit of the tree of the knowledge of good and evil.
- So, Lucifer's final words connected God with *"...knowing good and evil."* That statement was a straight-out lie that some people still believe today. God knows only good. He has never experienced evil in any way, nor does he have anything to do with evil. The ones who had known both good and evil were Lucifer and his followers.

Lucifer's words to Eve were full of deception. [Deception is nothingness, no reality at all.] Up until when Lucifer spoke to her, she and Adam had only heard God's words (the reality), so she had no concept of what a deception was. For all she knew, the serpent she saw might have been a messenger from God—God might have changed His mind. Eve was an innocent with no experience with sound waves other than God's.

(3) Adam also ate the fruit when Eve gave it to him. Lucifer had placed Adam in the situation where the man had to choose between

two commandments that God had already given him: to be fruitful and multiply to fill the earth[2] (for which he needed Eve) or to avoid eating the fruit of the tree of the knowledge of good and evil.[3] Adam had also proclaimed that Eve was part of his flesh since God created her from his rib.[4] So, Adam's dilemma was whether to forsake Eve or eat the forbidden fruit. Either way, he would disobey God, so he chose Eve—he didn't abandon her. Thus, the two humans fell for Lucifer's scam. Because they lacked experience, they had no understanding of what was happening. They didn't independently eat the fruit and didn't understand that Lucifer was manipulating them through the serpent. By today's legal standards, Adam and Eve were victims of a crime. And, to put Lucifer's crime in scientific terms, he generated sound waves that had the exact opposite amplitude as to what God had spoken. Lucifer's sound waves became destructive interference, negating what God had said, and Lucifer replaced God's words with nothingness. Lucifer introduced the humans to the knowledge of evil, an introduction that would change their lives and life on Earth.

(4) The Genesis account of the interchange between Lucifer and Eve was written by a man intent on laying a foundation for a religion focused on appeasing God. The writer phrased God's response to make it seem like God was punishing all three participants for their great sin against Him. But that wasn't it at all. God alone knew the realities of what had happened and the consequences of Lucifer's scam would be. Those consequences weren't going to happen because of any choice that God had made or would make. They were going to happen because of Lucifer, Eve, and Adam's choices. Those choices began the battle between God's Somethingness and Lucifer's nothingness on the Earth. While God's words to Adam and Eve described what evil and nothingness would do in their lives,[5] His words to Lucifer emphasized that he (Lucifer) hadn't accomplished his goals of ruling over the Earth and destroying the human race. On the contrary, Lucifer would continue slithering (instead of soaring) and begin engaging in a war with Eve and her seed.[6] Despite Lucifer's attempts, the humans would

eventually prevail over him. The conflicts between the human race and Lucifer's kingdom, Somethingness and nothingness, and good and evil were officially on.

(5) In Genesis 3:22-24 KJV, the writer explained that God banished Adam and Eve from the Garden of Eden because the humans had become like Him and knew good and evil. The writer didn't understand who God is—the One who generates good and doesn't know evil—and presumed that God didn't want the humans to eat of the tree of life and live forever. Again, the writer held God up as being offended and implied the need of the humans to appease an angry God. That wasn't it at all—Adam and Eve chose death by eating the forbidden fruit. God wasn't mad at them because He doesn't generate anger. By acting on Lucifer's deception, the humans moved out of the protection God offered them and into the realm that Lucifer was trying to control. That move was a literal move out of the Garden of Eden.

(6) Why didn't Lucifer's "win" cause God to forsake His Creation? The answer rests with sound waves. Before Lucifer and his followers decided to migrate from Infinity/Eternity to the Universe, there was no destructive interference against God's sound waves. So, Lucifer and his followers decided on their own and without interference. Once they decided to migrate, Lucifer and his followers removed themselves from God's sound waves to the silent Universe. What occurred in the Garden of Eden had a different set-up. Lucifer's deceptions negated what God had said to the humans—the lies didn't eliminate God's sound waves but canceled the amplitude of God's words. So, God's problem wasn't with Adam and Eve because they weren't the source of the deceptions. God's problem was how to deal with Lucifer's destructive interference so that the humans could hear Him. There was no reason for God to forsake the good He had created. Beyond that reality, God had a plan to work with the human race to eliminate Lucifer's destructive interference.

(7) In John 10:10 NASB, Christ said, *"The thief comes only to steal, kill, and destroy; I came that they may have life, and have it abundantly."* Through manipulation, Lucifer stole so much from the human race. He set them on the path to physical death and destroyed their open relationship with God. Lucifer made way for nothingness and dark energy to invade the Earth with his scam, bringing death and destruction with them. We need to understand the reality that wherever we see loss, death, and destruction, they are coming from the hand of Lucifer, not from the hand of God. Also, God does not control Lucifer. God will only intervene when humans look to Him for help and when they are willing to listen to His wisdom. As 2 Chronicles 16:9 NASB says, *"For the eyes of the Lord move to and fro throughout the earth that He may strongly support those whose heart is completely His."* God has not forsaken anyone, but He also will not force His way into anyone's life. He can offer to help, but we have the choice of accepting or not accepting that help.

Chapter Notes

[1] Genesis 2:16-17 KJV
[2] Genesis 1:28 KJV
[3] Genesis 2:16-17 KJV
[4] Genesis 2:24 KJV
[5] Genesis 3:16-19 KJV
[6] Genesis 3:14-15 KJV

Remember

[a] Chapter 5, Mini-Essay

Mini-Essay

Perhaps we should replace the word "sin" with "deception." The problem in the Garden of Eden wasn't the humans. They were innocent because they lacked experience. By focusing on their actions and condemning them for disobeying God, we miss the reality of what happened. Besides, Adam and Eve had the right to choose. [God doesn't force or require anyone to choose Him.] Their choice didn't change God as He continued to generate light, love, life, and good. And, His vibrations (words) didn't cease, so He was still active and present—He did not forsake His new Creation. Lucifer's deceptions were what was standing between God and the humans because they destructively interfered with God's reality. Also, when we use the word "sin," we concentrate on what we (and others) are doing and not on who God is. As a result, some of us who believe in Christ call ourselves "sinners saved by grace," which isn't the reality. We all are beloved by God. He hasn't deserted anyone. Our challenge is to discover what deceptions we believe and how those deceptions influence our words and actions. Somehow, we need to move Lucifer's sound waves out of the way so that we can hear God.

Suggested Reading

Cook, G. (2020, January 14). *Does Consciousness Pervade the Universe?* https://www.scientificamerican.com/article/does-consciousness-pervade-the-universe/

Hunt, T. (2018, December 5). *The Hippies Were Right: It's All about Vibrations, Man!* https://blogs.scientificamerican.com/observations/the-hippies-were-right-its-all-about-vibrations-man/

Phase - Interference - Higher Physics Revision. (2021).
https://www.bbc.co.uk/bitesize/guides/z99kkqt/revision/1

8

THE BIRTHS OF RELIGION AND COMPETITION

Key Words: Adam, Eve, Cain, Abel, Offering, Religion, Competition, God, Killing, Sacrifice, Ritual, Anger, Self-defense, DNA, Music, Art, Species, Apocalypse, Religious practices, Contradiction, Benefit, The Law, Reason, Deceptions, Reality, Love, Prayer Closet, Immortality

The Model

(1) The relationship between Adam and Eve's first sons, Cain and Abel, showed that the establishment of religion and competition happened early on in human history. Let's look at what transpired between the two brothers in Genesis 4:1-15 NASB. Cain's offering was of the crops he had harvested from the ground, while Abel's offering was one or more animals he had butchered from his flock (gatherer vs. hunter). God had regard for Abel's offering and none for Cain's, who became angry at the situation. God's response to Cain was:

> *"Why are you so angry? And why has your countenance fallen? If you do well, will not your countenance be lifted up? And if you do not*

> *do well, sin is crouching at the door; and its desire is for you, but you must master it."*

After God talked to him, Cain spoke to Abel and killed him. When God confronted him about Abel, Cain tried to evade responsibility for his brother's death, but God knew the reality of the situation. God spoke that the consequence of Abel's killing was Cain would become a vagrant and a wanderer on the Earth. When Cain feared someone would take revenge and kill him, God's answer was that anyone who killed Cain would suffer seven-fold vengeance. Then He appointed a sign for Cain (the KJV says God set a mark upon him) that would keep other people from killing him.

(2) Cain and Abel's story seems straightforward, but we might gain some new perspectives if we look underneath the surface.

- First, we must remember that the writer of Genesis intended to establish the Jewish religion, which centered on animal sacrifice and grain offerings. As a result, his focus was on Abel's sacrifice as the cause of Cain's anger. Both Cain and Abel seemed to believe that they had to appease God through a religious ritual.
- Second, how did God have regard for Abel's offering and not for Cain's? Did God really engender competition between the brothers? Was there some outward sign, or was Abel just boasting about how much better his animal sacrifice was than Cain's grain offering?
- Third, the first thing God said to Cain was, "Why are you angry?" [If there was an outward sign that God didn't regard Cain's offering, maybe the reason was Cain's anger, not his offering.] God went on to say that Cain needed to master his rage. Religion concentrates on rituals designed to appease God, while God desires that we master whatever brings loss, death, and destruction into our lives.

- Fourth, the writer doesn't say why Cain rose up against Abel and what Abel was doing at the time. If we look at what Lamech (Cain's great-great-great-grandson) said to his wives in Genesis 4:23-24 NASB, there seems to be a distinct possibility that Cain killed Abel in self-defense. Listen to what Lamech said: "*... For I have killed a man for wounding me; and a boy for striking me; if Cain is avenged sevenfold, then Lamech seventy-sevenfold.*" Lamech compared his self-defense actions to Cain's actions, which supports the feasibility of Cain killing Abel while defending himself. Also bolstering this possibility is that the consequence of Cain killing Abel wasn't a death sentence. [Deuteronomy 19:4-7 NASB describes a provision of cities of refuge to be set aside for people who unintentionally killed other people.] Cain may not have meant to kill his brother, which would explain God's provision to keep him safe from retribution from other people.
- Fifth, in the KJV, the last part of Genesis 4:15 says, "*... And the Lord set a mark upon Cain, lest any finding him should kill him.*" If that mark was a change in Cain's DNA—maybe one that changed his skin tone, hair color, eye color, stature, etc.—then a new species of humans could have been formed.
- Sixth, Cain's descendants went beyond being just gatherers and hunters (perhaps due to a change in his DNA). In Genesis 4:21-22 KJV, two of Lamech's sons introduced music and art (working with brass and iron) to the human experience. Understanding how to express oneself through the fine arts can mediate the effects of anger.
- Seventh, Genesis 6:1-4 KJV talks about the sons of God taking the daughters of men as wives and having children with them. This distinction would make sense if Cain's descendants (perhaps with lighter hair, eyes, and skin) were a different species from the rest of Adam and Eve's lineage. These few verses in Genesis 6 open the possibility of several new species coming out of the

> intermarriages, as they specifically mention giants. [If there were giant humans, it makes sense that there were huge animals, too.] Scientists have found evidence of various human species, and this passage gives credence to their discoveries.

So, great diversity grew on the Earth as biological life bore fruit and multiplied. But, an apocalypse was coming, and it would change everything.

Chapter Notes

None

Remember

None

Mini-Essay

To understand what God thinks about religious practices, let's look at Isaiah 1:11-15 NASB, where He spoke through the prophet:

> *"What are your multiplied sacrifices to Me?" says the Lord. "I have had enough of burnt offerings of rams and the fat of fed cattle; And I take no pleasure in the blood of bulls, lambs or goats.*
>
> *"When you come to appear before Me, who requires of you this trampling of My courts? Bring your worthless offerings no longer. Incense is an abomination to Me. New moon and Sabbath, the calling of assemblies—I cannot endure iniquity and the solemn assembly.*

> *"I hate your new moon festivals and your appointed feasts, they have become a burden to Me; I am weary of bearing them. So when you spread out your hands in prayer, I will hide My eyes from you; Yes, even though you multiply prayers, I will not listen. Your hands are covered with blood."*

Talk about a significant disconnect: Moses initiated specific rituals for the Jewish religion, and Isaiah rebuked the Jews for those rituals. Is God double-minded?

Probably the best way to look at this contradiction is to understand that religious practices were to benefit the human race rather than God. In Mark 2:27 NASB, Christ said, "*... The Sabbath was made for man, and not man for the Sabbath.*" Since the Sabbath was part of the Law that Moses put together (while alone with God), we can extend Christ's statement to include all of the Law. God gave Moses instructions to build a way for people to remember God until such a time when there would be enough understanding of God to make the rules irrelevant. Later on, even the Apostle Paul knew that his knowledge of God was lacking as he wrote, "*For now we see in a mirror dimly, but then face to face; now I know in part, but then I will know fully just as I also have been fully known*" (1 Corinthians 13:12 NASB). Paul knew that God knew him thoroughly, but his knowledge of God was sketchy. Like Moses, Paul wrote down a new series of rules and rituals to establish a new religion. The main difference between the two codes was that sacrifices were no longer needed. Yet, Paul still set up Christ as the ultimate sacrifice for sin, which wasn't the purpose of Christ at all. (We'll talk about that purpose in Chapter 12.) So, the new religion focused on sin and appeasing God by accepting Christ and following the new rules.

Let's go back to Isaiah 1 NASB and look at verses 16-20. God had told the Jews what He hated in the previous verses, but He didn't leave them hanging. He went on to challenge them with the following words:

> *"Wash yourselves, make yourselves clean; remove the evil of your deeds from My sight. Cease to do evil, learn to do good; seek justice, reprove the ruthless, defend the orphan, plead for the widow.*
>
> *"Come now, and let us reason together," says the Lord, "Though your sins are as scarlet, they will be white as snow; Though they are red like crimson, they will be like wool.*
>
> *"If you consent and obey, you will eat the best of the lands; But if you refuse and rebel, you will be devoured by the sword." Truly, the mouth of the Lord has spoken.*

God wanted and still wants people to reason together with Him—when we reason with God, the written rules and rituals become unnecessary. God wants to explain things to us because deceptions will be swallowed up by His reality when we understand that reality. When those deceptions are gone, we will be able to obey God by loving Him and loving our neighbors as we love ourselves. Yes, reasoning with God takes time and effort in our prayer closets, but that work is necessary if we desire to experience immortality.

Suggested Reading

Plackett, B. (2021, January 24). *How many early human species existed on Earth?*
https://www.livescience.com/how-many-human-species.html

Sacrifice in Ancient Israel. (n.d.).
https://www.bibleodyssey.org/en/passages/related-articles/sacrifice-in-ancient-israel

Sacrifices and Offerings (Karbanot). (n.d.). https://www.jewishvirtuallibrary.org/sacrifices-and-offerings-karbanot

9

THE GREAT FLOOD AND THE SHATTERING OF TIME

Key Words: Great Flood, Noah, Ark, God, Extinction Event, Fountains of the Deep, Magnetic Field, Asteroid, Tsunami, Time, Dating Methods, Lucifer, Deception, Nothingness, Reality, Choice, Love, Energy, Moses, Humans, Promise, Descendants, Rain, High-Energy Cosmic Rays, Hydrogen, Electrons, Oxygen, Chemical Reaction, Turbulence, Atmosphere, Plasma, Vibrations, Apocalypse, Rainbow

The Model

(1) Many religions and cultures have the tradition of a worldwide flood that killed off the vast majority of plant and animal life; among them, the Genesis account of Noah is one of the best-known. The general take on Noah is: *After constructing the ark as God had instructed, Noah rounded two of every animal and loaded them and his family into an ark. After God closed the door behind them, the rain started and fell for forty days and forty nights. After the water subsided, Noah, his family, and the animals disembarked, and Noah offered up sacrifices to God. God responded with the sign of the rainbow and a promise never to flood the Earth again. Of course,*

God caused the rain to happen in the first place because the people on Earth were so wicked. This tradition of Noah is religious (with the implication again that humans need to appease God). It waters down what the Bible says about the events during this extinction event. Much more than forty days of rain occurred—as the writer in Genesis put it, " *... the same day were all the fountains of the great deep broken up ...* ".[1] If we skim over those words, we could assume that the fountains were streams of water directly underneath the surface of the Earth (the crust). We should remember that the magma in the mantle and in the outer core is liquid, too. All the fountains of the great deep may have included what erupted from the mantle and outer core.

(2) If we put together what the Bible relates about the days of Noah with the theories and findings that scientists have come up with concerning extinction events, the details of what happened during the Great Flood becomes clearer. As we look at those details, we're going to blank out the time frames that scientists have used and look only at the events and circumstances. Here are the relevant discoveries:

- Scientists theorize that, at one point, Earth had just one landmass surrounded by one body of water. The Bible concurs with this assessment as Genesis 1:6 NASB says that God gathered all the water on our planet into one place and then made the dry land appear. So, until the time of Noah, only one continent and one ocean may have existed on Earth.
- Scientists have found evidence that the Earth's magnetic field was once a tiny fraction of what it is today, with the field reversed from its present position.
- More evidence suggests that numerous asteroids plummeted the planets closest to the Sun.
- Scientists believe an asteroid hit the Earth and caused a tsunami almost a mile high that killed all dinosaurs. Recent evidence shows unusual blobs and giant mountains within the Earth's

mantle; the scientists connect at least one blob with the asteroid hit.

- Finally, scientists have found that the Earth may have a previously hidden inner-inner core which suggests that the Earth had a second extreme cooling event that has been unknown to scientists. The crystals in the inner core point in a north-south direction, while the newly-found core has crystals pointing in an east-west direction.

What if these occurrences connect to what happened during the Great Flood? What if there was only one apocalyptic extinction event rather than multiple events over billions of years?

(3) The Bible hints that the Great Flood was more than just a worldwide flood. In Genesis 5, the writer gave a run-down of the male side of Noah's genealogy. Most of the men lived between 800 and 1000 years. Genesis 11 shows that, after the flood, lifespans first decreased to about half of the original spans and then to half the second amount. This decline in lifespans raises the possibility that the Great Flood shattered the time system that Noah's ancestors had known and brought to pass a new time system. A change like that would explain the discrepancy between the lifespans before and after the Great Flood. With this in mind, here's a possible scenario of what happened:

- When the humans and animals entered the ark, the Earth was composed of one landmass and one body of water. God secured Noah and his boat-mates in the ark right before the asteroids began bombarding the solar system. [Dark energy may have fueled the bombardment.]
- The force of the continuous hits shattered the crust and mantle of the Earth, which caused unending earthquakes and volcanoes, which produced tsunami after tsunami, which flooded the entire landmass. The Earth turned inside-out and outside-in multiple times. Of course, this wiped out the vast majority of life forms.

- The unrelenting force of the asteroid hits sent the Earth tumbling in space and spinning out of control, thus almost destroying its magnetic field. At that point, time ceased to be what it was.
- As the bombardment ended, the Earth eventually wobbled itself into a new orbit farther away from the Sun, which led to a significant cooling and a new measurement of time with longer years and possibly longer days. Until the Earth settled into its new orbit, time didn't exist.
- So, Noah and his boat-mates spent part of their lives without the consistency of time, howbeit unknowingly. [Humans don't feel the movement of the rotations and revolutions of our planet, solar system, and galaxy.] Inside the ark, the humans and animals had no reference to keeping track of the passage of time.
- When the Great Flood waters receded, the Earth's one landmass and one continuous ocean were broken up into many landmasses (from small islands to massive continents) and many oceans that remained connected but were much farther apart.
- As the constant earthquakes and volcanoes calmed down, the rock layers in the crust and mantle were jumbled from their original states. Carcasses of animals and humans, along with human-made artifacts, had been either swallowed up by the heaving Earth or hurled into the raging waters of the tsunamis. Thus, evidence of both animal and human life was relentlessly scattered and buried in varying depths beneath the new surface of the Earth.

If this scenario is close to reality, the dating methods so carefully designed by scientists would be moot. How can one date objects that existed in the original time frame, in the period with no time, and in the newly-formed time frame? Being accurate would be a challenge, if not impossible. We cannot assume that time remained stable during this extinction event. If the circumstances were enough to wipe out the majority of life on the Earth, then considering the effects of those same circumstances on time is reasonable and prudent.

(4) The writer of the Genesis account put the "blame" of the loss of life on God and lamented that the wickedness of the human race caused God to grieve and regret that He had made the human race.[2] However, this was the early understanding. In Matthew 24:38-39 KJV, Christ clarified that before the Great Flood, people participated in everyday activities, not in any great, unheard-of wickedness. And, remember that Christ's statement in John 10:10 NASB clarified that it is the thief who comes to steal, kill, and destroy. God doesn't destroy—Lucifer and his followers are the ones who bring destruction through dark energy and nothingness. [To believe that God is the one who destroys is to believe that God has given up on His goals and His work.] This understanding of the destruction coming through Lucifer, though, brings more questions: Why didn't God intervene and stop this apocalyptic event from happening? If God had created all of the matter in the Universe, why couldn't He have stopped the wholesale destruction of the Great Flood extinction event?

(5) To find the answers to these questions, we need to look at the natures of God and Lucifer. Remember,[a] we've already noted that God generates light, life, love, and other good things, but Christ adds dimension to God's character with the words he spoke in Revelation 3:20 KJV. Here's what he said: "*Behold I stand at the door and knock: if any (human) hear my voice, and open the door, I will come into (them), and will sup with (them), and (they) with me.*" God extends the opportunity to know Him to all of us but doesn't demand that we take that opportunity. The choice is entirely ours. Now, let's look at how Lucifer acts—I Peter 5:8 NASB is specific about him: "*Be of sober spirit, be on the alert. Your adversary, the devil, prowls around like a roaring lion, seeking someone to devour.*" This description personifies dark energy and its determination to destroy, to tear things apart. Lucifer doesn't wait for an invitation to enter our lives with his destruction; he barges right in and wreaks havoc (almost complete devastation in the case of the Great Flood extinction event). Remember,[b] one of Lucifer's greatest weapons

is the nothingness of deception. He is aggressive with his destructive interference to keep people from connecting with God. In the days before the Great Flood, Noah was the only one who pressed through the destructive interference, and he did that by diligently seeking God,[3] which was not an easy thing to do.

(6) So, does God hate the people who don't connect to Him? No. God understands that Lucifer's deceptions plague the human race; the deceptions result in a lack of knowledge of reality, which leads to death.[4] Per Christ's words on the cross, "*Father, forgive them; for they do not know what they are doing.*" (Luke 23:34 NASB), God doesn't hold our lack of knowledge and understanding against us. His whole purpose is to bring all of us to a point where we can understand the difference between deception and reality. That's when we can make a legitimate choice between God and Lucifer. For many people, such as the ones who perished in the Great Flood, the knowledge of reality won't happen until the Great White Throne Judgment,[5] which takes place sometime after the return of Christ (more about this in Chapter 15). We can rest assured that God's love for the world He created will not fail[6] in the end as He works diligently with the human race to help them understand reality.

(7) Now, let's go back to Noah, his family, and all the creatures in the ark. Even for those who believe the ark is a myth, the reality is that some life forms on the Earth survived this extreme extinction event. Somehow, someway, a few human and animal lives were kept safe. Because water covered the Earth, the theory that a boat of some sort was involved sounds reasonable; however, the tsunamis that accompanied the asteroid bombardment should have swamped the ark. How did it sail through the onslaught without being destroyed? Perhaps, the same energy that parted the Red Sea so that Moses and the Hebrews could escape Pharaoh[7] kept the waters of the tsunamis away from the ark. Regardless of how it happened, the survival of the ark during the massive devastation of this extinction event speaks of God's faithfulness to

humans who diligently seek Him. Again, God desires to be faithful to all humans and all life on the Earth, but He will never force the issue.

(8) As mentioned earlier, other religions and cultures have accounts of the Great Flood (some written well before the Genesis writer told Noah's story). The difference between those "myths" and the Bible is that Noah's genealogy (before and after the extinction event) was recorded to such an extent that we can be reasonably sure that Noah was a real person. [The biblical authors excelled in recording the genealogies of people in the Hebrew nation along with their population numbers.] Another difference between Noah's story and many other accounts is that he, his family, and their descendants were real humans, not the mythical children of deities and humans. But, what matters most is the human race and some animal life survived this cataclysmic extinction event. [We can be happy about that because, without that survival, none of us would be here.] Also important: after the bombardment of the asteroids ceased and the waters receded, God was able to make a promise to Noah that the destruction of all flesh would not happen again.[8] The first thing Noah did when he, his family, and the animals exited the ark was to build an altar and burn offerings to God.[9] Although the religious rituals survived the Great Flood, Noah was diligent in honoring God and seeking Him through those rituals. As a result, God promised that there would never be a similar extinction event again.

Chapter Notes

[1] Genesis 7:11 KJV

[2] Genesis 6:5-7 KJV

[3] Hebrews 11:6-7 KJV

[4] Hosea 4:6 KJV

[5] Revelation 20:11-15; I Corinthians 3:11-15 KJV

[6] I Corinthians 13:8 NASB

[7] Exodus 14 NASB
[8] Genesis 9:15 NASB
[9] Genesis 8:20-21 NASB

Remember

[a] Chapter 2, Paragraphs 3, 4, 5, 6
[b] Chapter 7, Paragraph 2

Mini-Essay

Although the Great Flood had more to do with asteroids and tsunamis than rain, let's talk about how water comes down from the sky and then talk about climate change. First, the rain: High-energy cosmic rays (carrying nuclei, mostly hydrogen) continuously enter the Earth's atmosphere, where free electrons are abundant. As the rays penetrate the oxygen-rich atmosphere, the combination of the high level of energy, the hydrogen nuclei, the electrons, and the oxygen can produce chemical reactions at times. Those reactions result in water (rain) and explosions (lightning and thunder). Under certain circumstances, the chemical reactions can have tremendous turbulence, producing more chemical reactions. Those secondary reactions can cause other compounds (like carbon dioxide) to break apart. The updrafts created by the turbulence can carry the remnants of the compounds out of the Earth's atmosphere, which releases 90 tonnes of plasma per day. [If this description of the creation of water in our atmosphere is close to reality, the Earth seems to have an efficient system for dealing with any harmful compounds that we produce.]

Now, about climate change: the problem may not primarily be the greenhouse gases in our atmosphere retaining heat. The problem could center on the amount of heat we produce through the vibrations of human activities and loud human voices. Think about the sounds that pound the Earth's atmosphere daily—those sounds generate heat. Perhaps, the answer to climate change isn't only about cutting back on carbon dioxide but also about lowering the volume of vibrations (voices and sounds).

There may be a reason for many of us being interested in a possible apocalypse that would end our world. We all have descended from the people who lived through that Great Flood. Those survivors suffered trauma that might have inscribed itself genetically within their DNA/RNA. That memory may have passed from generation to generation through all the millennia. So, the fascination and fear of the world ending may have their roots in the Great Flood. If the Genesis account is close to the reality of what happened, we should remember that God promised Noah and his family that a similar event would never happen again. In the Genesis 9:8-17 NASB account of that new covenant, God extended the promise to Noah's descendants (including all of us) and set up the system of rainbows as a token of that declaration. Looking at the rainbows scientifically, God may have adjusted the way the Sun and Earth interacted to prevent such an onslaught of asteroids happening again. The energy that saved Noah and his family in the ark may be protecting the Earth now.

Suggested Reading

ESA Science & Technology - The curious case of Earth's leaking atmosphere. (2016, July 7).
https://sci.esa.int/web/cluster/-/58028-the-curious-case-of-earth-s-leaking-atmosphere

Geggel, L. (2019, January 7). *Dinosaur-Killing Asteroid Triggered Mile-High Tsunami That Spread Through Earth's Oceans.*

https://www.livescience.com/64426-dinosaur-killing-asteroid-caused-giant-tsunami.html

Ghose, T. (2018, February 23). *Facts About Pangaea, Ancient Super-continent.*
https://www.livescience.com/38218-facts-about-pangaea.html

Henriques, M. (2019, March 26). *Can the legacy of trauma be passed down the generations?*
https://www.bbc.com/future/article/20190326-what-is-epigenetics

How long would you have to yell to heat a cup of coffee? (2022).
https://www.physicscentral.com/explore/poster-coffee.cfm

Howell, E. (2018, May 11). *What Are Cosmic Rays?*
https://www.space.com/32644-cosmic-rays.html

The Ionosphere | Center for Science Education. (2022).
https://scied.ucar.edu/learning-zone/atmosphere/ionosphere

Jackson, C. (2019, March 2). *What Happens When Hydrogen & Oxygen Combine?*
https://sciencing.com/happens-hydrogen-oxygen-combine-8515474.html

National Geographic Society. (2019, April 4). *Magma.*
https://www.nationalgeographic.org/encyclopedia/magma/

Our planet may have been last pummeled by asteroid impacts longer ago than previously thought, explaining why life began to form almost 4 billion years ago. (2019, August 15).

https://astronomy.com/news/2019/08/new-date-for-late-heavy-bombardment-may-change-lifes-timeline-on-earth

Powell, C. S. (2019, April 24). *Deep inside Earth, scientists find weird blobs and mountains taller than Mount Everest.* https://www.nbcnews.com/mach/science/deep-inside-earth-scientists-find-weird-blobs-mountains-taller-mount-ncna997111

Spry, J. (2021, March 11). *New study reveals Earth's hidden inner core and suggests a mystery cooling event.* https://www.syfy.com/syfywire/new-study-reveals-earths-hidden-inner-core

Thompson, H. (2021, July 29). *Dinosaur-killing asteroid may have made Earth's largest ripple marks.* https://www.sciencenews.org/article/dinosaur-killing-asteroid-chicxulub-earth-largest-ripple-marks-tsunami

Whitt, K. K. (2021, February 21). *Upheaval and extinctions linked to magnetic reversal 42,000 years ago | Earth | EarthSky.* https://earthsky.org/earth/magnetic-field-reversal-42000-years-ago-neanderthal-extinction/

10

OLD TESTAMENT BREAKTHROUGHS

Key Words: Old Testament, Hebrews, God, Abraham, Isaac, Sacrifice, Lucifer, Christ, Burnt Offerings, Religious Practices, Moses, The Law, Ahaz, Hezekiah, Hypocrisy, David, Bathsheba, Uriah, Nathan, Judgments, Sin, DNA, Appease, Truth, Heart, Reality, Vanity, Glass Ceiling, Eternity, Time, Babel, Sodom, Volcano, Gomorrah, Dark Energy, Lot, Esau, Jacob, Rebekah, Birthright, Blessing, Burning Bush, Holy, Ground, Holy Spirit

The Model

(1) The Old Testament of the Bible is a detailed history of the establishment of the Hebrew nation and its development. For the most part, the central concept of appeasing God under-girded how the writers recorded the events in the Old Testament. According to those writers, when the Hebrews followed the rules, God blessed them, and when they didn't follow the rules, God moved against them. So, God was the highest being who ruled the world, and He exercised vengeance against those who disobeyed Him. The problem with this concept was the emphasis on what humans do instead of who God is. Yet, there were

breakthrough moments when glimmers of the reality of God appeared within the recorded accounts.

(2) One such moment occurred when God asked Abraham to sacrifice Isaac, as related in Genesis 22. The writer framed the incident as God testing Abraham, and Abraham passing the test because he was willing to sacrifice the son God had promised him. Again, the context of the written tale was the need to appease God. But what if that wasn't it at all? What if God took a stand against one of the most horrific practices that humans participated in at that time? What if God spoke a resounding "No!" to human sacrifice? In Abraham's day, people sacrificing other people to appease their gods was common. What if God asked Abraham to sacrifice Isaac because He knew Abraham would stop when He told him to stop? Remember,[a] God is and generates life. He has nothing to do with death. The purveyor of death is Lucifer—Christ emphasized this reality by stating the devil was a murderer from the beginning.[1] When humans sacrificed other humans to appease a god, the one they were "appeasing" was Lucifer; they played directly into his hands. So, because God didn't want the human sacrifice to be associated with the Hebrews worshipping Him, He made a definitive statement when He told Abraham to stop just as the man picked up the knife to sacrifice his son. God directed the focus of sacrifice away from humans to animals by providing an animal substitute for Isaac. When thinking about this, remember[b] that during the time of Isaiah, God stated that He hated the burnt offerings of animals.[2] This incident with Abraham and Isaac was the first step away from the practice of offering burnt sacrifices to appease God and away from the concept of having to appease God.

(3) God worked with Moses many years later to set up a burnt offerings system and other religious practices. Again, wasn't that system a giant step backward and a contradiction to what God later said to Isaiah? Why would God ensure the Hebrews would work to appease Him if He didn't need the appeasement? The answers lie in the human

experience, not the nature of God. The Hebrews needed to be engaged in practices that helped them remember that God existed. The incident with the golden calf (Exodus 32 NASB) powerfully illustrated that need. Moses had left the Hebrews to be with God on Mount Sinai for forty days and forty nights, so the people were getting restless and impatient. Let's look at what the Hebrews said to Aaron about Moses's absence in Exodus 32:1 NASB: "*Come, make us a god who will go before us; as for this Moses, the man who brought us up from the land of Egypt, we do not know what has become of him.*" For all the Hebrews knew, a wild animal could have killed Moses—they were justified for being uneasy. That they reverted to the practice of creating a god to appease is understandable. After their spectacular liberation from Egypt, the Hebrews needed to be engaged to maintain hope when the deliverance morphed into an unknown future.

(4) God knew this, so He worked out a system with Moses to fully engage the Hebrews in religious practices that would give them hope. This system, the Law, served as a beginning placeholder as God endeavored to equip the human race with the knowledge, understanding, and wisdom to discern between deception and reality. And, when the Hebrews inevitably walked into nothingness and dark energy, the Law was there to pull them back. An example of this getting lost and coming back is the story of King Ahaz and his son, King Hezekiah, as told in II Chronicles 28-32 NASB. Ahaz completely uprooted any worship of God and threw himself and Judah into idol worship which included sacrificing his children. After Ahaz died, Hezekiah ultimately returned to Moses's system. [The Hebrews didn't have an anchor without the system.] So, God was willing to endure the religious practices as substitutes for the end goal He was working towards: a one-on-one relationship with each person that would enable that person to overcome deception, dark energy, and nothingness.

(5) The religious practices that formed the Law had a significant lack that hindered the full manifestation of God's knowledge, understanding, and wisdom to individuals. The practices were based on what people did and not on who they were. Focusing on the outward appearances of obeying the Law often led to hypocrisy because judging others was much easier for people to do than dealing with what goes on in their own inner lives. King David and Bathsheba's story (as related in II Samuel 11-12 NASB) makes this apparent. After David committed adultery with Bathsheba, she conceived and let David know. David brought her husband Uriah back from the battlefield and encouraged him to go home to sleep with his wife. But, being super-loyal to David, Uriah refused, so David sent him to die on the battlefield. After Bathsheba mourned her husband, David married her. The writer of II Samuel was specific that David did evil in the sight of the Lord.[3] David's adultery and his manipulation to assure Uriah's death broke the Law.

(6) David had no inner knowledge of the evil he had done to Uriah. He possibly could have justified his actions that he tried his best to get Uriah to have relations with Bathsheba and that Uriah was so loyal that he probably would think it was an honor to die for his king. At that point, God had no way of speaking directly to David because he didn't see anything wrong with what he did. In the end, God sent the prophet Nathan to confront David, and the king came face to face with his hypocrisy and the evil of his actions. Nathan told David a story similar to what he had done; David became very angry with the tale's villain and called for restitution for the victim. He had no concept that he was the villain until Nathan spoke with blunt certainty, "*You are the man!*"[4]. The prophet explained to David the depth of his evil and the consequences of his choice: "*... the sword will never depart from your house ...*" .[5] [Remember,[c] the Universe is full of circles and ellipses that rotate and revolve. Similarly, our words and actions circle back to us. Some people call it karma—Christ put it in blunt terms: "*For in the way you judge, you will be judged; and by your standard of measure, it will*

be measured to you."[6] As David learned, the harsh judgments we make towards other people will boomerang back to us.]

(7) After Nathan spoke to him, David spent some time thinking and writing about what he had done. In Psalm 51 NASB, we see how David was processing the reality of the depravity of his actions towards Uriah. However, he missed several salient points but understood a fundamental reality and prayed an effective prayer. The first group of salient points appears in Psalm 51:4-5 NASB, where David claimed that his sin was only against God and that he was brought forth in iniquity from the moment of conception. David's sin didn't affect God; nothing we humans do changes anything God does. He always generates light, life, love, good, energy, and vibrations. We alone decide whether we will benefit from all that God does. In this case, David's sin was against Uriah, who bore the full brunt of David's adultery and manipulation. Then, as far as David being brought forth in iniquity goes, there is a partial reality to that understanding because we do inherit "tendencies" through our DNA. However, his sin against Uriah came from deliberate actions that David had chosen to do. Those actions had nothing to do with his conception—David knew what he was doing and didn't accidentally fall into committing adultery and manipulating a man's death.

(8) The second grouping of the salient points David missed occurs in Psalm 51:16-19 NASB. He understood that the results of his wrongdoing have nothing to do with appeasing God; in fact, he realized that God doesn't delight and isn't pleased with sacrifices and burnt offerings. David grasped that God is looking for broken spirits and contrite hearts. But then, David immediately turned around and asked God to do good so that He will once again delight in the sacrifices and burnt offerings of the Hebrews. David didn't realize that God didn't need appeasement but that whatever drove David to adultery and murder had to be dealt with by David with God's help. Otherwise, he would

continue to dwell in the vicious circles of karma until his death (which was what happened). So often, humans try to make God responsible for the choices that bring judgments back to them.

(9) Amid the misunderstandings in his song, David did understand one reality: God desires truth in the innermost being, the hidden part where God makes known His wisdom (Psalm 51:6 NASB). One thing the religious practices of the Law could never do was change a person on the inside. Everything involved with the Law was focused on what people do and not on who they and God are. David saw the lack in his own life and asked God to make changes on his inside being: "*Create in me a clean heart, O God, and renew a steadfast spirit within me. Do not cast me away from Your presence and do not take Your Holy Spirit from me.*" (Psalm 51:10-11 NASB). David needed a change within the depths of his being, a change that the Law could not facilitate. This prayer of David made room for promises that God gave to Jeremiah and Ezekiel many years later. These pledges pertain directly to David's prayer:

- Jeremiah 31:33-34 NASB: "*But this is the covenant which I will make with the house of Israel after those days," declares the Lord, "I will put My law within them and on their heart I will write it ... They will not teach again, each man his neighbor, and each man his brother, saying, 'Know the Lord,' for they will all know me ...*".
- Ezekiel 36:26-27 NASB: "*Moreover, I will give you a new heart and put a new spirit within you ... I will put My Spirit within you and cause you to walk in My statues ...*".

God's plan is to replace the Law with a reality that gives humans access to Him. If they choose that access, the way opens up for God to work with them to bring about changes in their inner beings. When that happens, the need for the placeholder of the Law no longer exists.

(10) Perhaps, one of the best explanations of why the Law was only a placeholder came from the writer of the Book of Ecclesiastes. The words of one of the wisest kings in Israel reeked of pessimism as he lamented that life was futile and full of vanity. He concluded that everyone dies and goes to one place, so what was the use of being wealthy or wise or righteous—bad things happen to the righteous, and good things happen to the wicked. In Ecclesiastes 9:2-3 NASB, the writer bemoaned, "*It is the same for all. There is one fate for the righteous and for the wicked ... that there is one fate for all men ... Afterwards they go to the dead.*" (Notice that the writer had no concept of Heaven and Hell but believed that everyone went to the same "place" after dying.) The Law and its religious practices did nothing to change anything. From the sentiments expressed in Ecclesiastes, we can infer that the writer must have been full of anticipation as he began his quest for wisdom that he was willing to work so hard to attain. Instead of reaching the stars, he ran into a glass ceiling that sent him barreling backward to realize all his work had been for nothing.

(11) We can find one of the best-known passages of the Bible in Ecclesiastes 3:1-8 NASB:

> "*There is an appointed time for everything. And there is a time for every event under heaven—A time to give birth and a time to die ... A time to kill and a time to heal ... A time to weep and a time to laugh ... A time to love and a time to hate; A time for war and a time for peace ...*".

The writer believed that humans had no control over time or what happened to them. A few verses later, the writer hones in on what goes to the very heart of the glass ceiling. The last statement in verse 11 is, "*He (God) has also set eternity in their heart, yet so that man will not find out the work which God has done from the beginning even to the end.*" Remember[d] when we talked about our DNA being eternal? Considering that the same DNA makes each of us a unique individual, we could say that our DNA is our heart, and we have eternity in our hearts. But, to the writer of Ecclesiastes, that eternity made no difference. At that time, people were so caught up in the framework of time that they couldn't comprehend the work God had done, was doing, and would do. We could say that time was the glass ceiling that kept people from finding the eternal God.

(12) The Law, the placeholder, could not shatter the glass ceiling of time. Going back to David, his inability to see eternity, to see God's work from before until the end, kept the man from being able to overcome the urge to commit adultery with Bathsheba and the determination to arrange Uriah's death on the battlefield. David's prayer and God's later promises looked ahead to the point when the human heart—the human DNA—would undergo a change that would shatter the glass ceiling of time.

Chapter Notes

[1] John 8:44 KJV
[2] Isaiah 1:11 NASB
[3] II Samuel 11:27 NASB
[4] II Samuel 12:7 NASB
[5] II Samuel 12:10 NASB
[6] Matthew 7:2 NASB
[7] Ecclesiastes 1:2 NASB

Remember

[a] Chapter 2, Paragraph 4
[b] Chapter 8, Mini-Essay

[c] Chapter 5, Paragraph 3
[d] Chapter 6, Paragraph 4

Mini-Essay

Let's look at four Old Testament incidents from the perspective of this new model. First, the Tower of Babel: The account in Genesis 11:1-9 NASB again holds God responsible for what happened with Babel when different languages first appeared. The writer implies God didn't want the human race to reach for the stars, so He jumbled their single language into multiple languages and scattered the people throughout the world. A more realistic explanation is a group of people got together to complete a massive project, but competition soon arose. Factions developed and created secret codes that soon became official languages. Since there was so much space on the Earth's surface, the different groups split up to pursue their own goals. Since God doesn't control people nor demand His way, the culprit of competition makes sense. And in today's world, the competition is so intense that factions within groups abound. Any entity or person wanting to control the world will never succeed (conspiracy theorists can give it a rest).

Second, the destruction of Sodom and Gomorrah, as related in Genesis 18:16-19:29 NASB: God saw that a volcano—caused by dark energy—was about to erupt and wipe out the cities of a valley, which included Sodom and Gomorrah. Looking for a way to save the cities, He told Abraham about the destruction because He knew the man would intercede. Abraham prayed that God would stop the carnage for the sake of ten "righteous" people (those who would listen to God). However, the inhabitants were more interested in pleasure and partying than listening to God (their legitimate choice). No one but Lot, his wife, and their two daughters escaped Sodom. When Lot asked that they be sent to another city in the valley rather than into the mountains, God was able to limit the destruction in the valley so that particular city would be safe. Of course, people running from a volcano don't want to look towards the eruption because the gases would suffocate them. Lot's wife suffered such a fate as she looked back and hot ash (that eventually cooled into a pillar) covered her.

Third, the story of Esau and Jacob, the twin sons of Isaac and Rebekah (Genesis 25:19-34, Genesis 27-28, 32-33 NASB): Because the Apostle Paul and other biblical writers knew nothing about DNA, they made some assumptions about Jacob taking Esau's birthright and blessing. The consensus in the Bible was that God hated Esau and loved Jacob (Romans 9:10-13 NASB). But that's not what happened because God cannot hate anyone. To understand what happened with the twins, let's keep in mind what Psalm 139:13-16 NASB tells us: God knows us before we develop in our mothers' wombs. God read the DNA of the boys when they were conceived and knew their personalities and other genetic traits from that point on. When Rebekah asked God to explain the war in her womb, God told her about her boys' DNA configurations. The younger one was a cunning fighter and would "rule" over the older one, who was impulsive and led by what he felt moment by moment. For example, when Esau was hungry, and Jacob convinced him to sell his birthright for some stew—Esau had no thought about the consequences of that action. One of Esau's positive traits was that his anger flared but would die just as quickly. Rebekah said as much when she and Isaac sent Jacob away after Jacob conned Isaac for Esau's blessing: "*Stay with him (Laban, her brother) a few days, until your brother's fury subsides and he forgets what you to him. ...*" (Genesis 27:44-45 NASB). Rebekah knew her boys well. Also, Jacob was inclined to spend time diligently seeking God, while Esau didn't make that effort a priority. As it turned out, the boys both prospered and eventually reconciled with no hard feelings. So, their positions concerning the founding of the two nations God spoke of to Rebekah (Genesis 25:23 NASB) came because of their personalities and choices and not because of God's supposed hatred.

Fourth, Moses and the burning bush: Exodus 3 NASB relates how Moses became the leader of the Hebrew nation. One day, he came upon a burning bush that wasn't consumed by the flames. When he moved closer to check it out, God spoke from the bush and told Moses to remove his sandals because the man was standing on holy ground. If we apply what we've learned so far to this account, this incident takes on a whole new meaning. The burning bush was God's energy (the Holy Spirit) which doesn't consume matter but creates it. And that same energy infused the ground Moses stood on and made it holy (full of what God generates). When Moses took off his sandals, God's power infused his body. That energy enabled Moses to perform the miracles that he did, including the parting of the waters of the Red Sea.

11

UNDERSTANDING CHRIST

Key Words: God, Promise, Mary, Joseph, Christ, Holy Spirit, Gabriel, DNA, Word of God, Universe, Vibrations, Son of Man, Son of God, Ministry, Religious Leaders, Conflict, Prayer Closet, Faith, Love, Judge, Rich Man, Lazarus, Abraham, Religious Practices, Hypocrites, Miracles, Realities, Healing, Power, Wisdom, Obedience, Immortal, Emptying, Death, Helper, Science, Creation, Trinity

The Model

(1) About six hundred years passed from when God's promises were given to Jeremiah and Ezekiel until He took the first step to fulfill those promises. Why did it take so long? The simple reason is that God had to wait for the right set of circumstances to happen before His work could begin. Since the lynchpin of His plan involved a human baby born to a virgin mother, He had to check off several boxes. The proper young woman had to exist—she had to be able to understand what God was asking her to do and the choice she had. Then, the right man also had to be available because God wouldn't leave the young virgin and her baby unprotected. This man had to be honorable, willing to listen to God, and capable of choosing to follow God's instructions. Finally,

the young virgin and the man had to be in a relationship. Mary[1] and Joseph[2] met all of God's requirements. God's patience in waiting for everything to line up paid off big time.

(2) Christ's conception came about when the Holy Spirit overshadowed Mary. In Luke 1:26-38 NASB, as the angel Gabriel spoke to her about God's goal for her to become the mother of His Son, Mary asked a pertinent question, "*How can this be, since I am a virgin?*". [3] Gabriel answered that the power of the Most High would overshadow Mary as the Holy Spirit would come upon her,[4] which was similar to the method God used to create matter within the boundaries of the Universe.[5] [The power of the Most High included the general energy of the Holy Spirit and everything else God generates: vibrations, light, love, life, and good.] When this overshadowing happened, two of Mary's gametes were fused to give Christ a complete human genome. As the melding occurred, God's power "cleaned" out everything in Christ's genome that would lead to deception and death (remember this). Because his DNA was "purified" by God's power, Christ was the Son of God and Mary's Son. Also, because a Y chromosome wasn't involved in his conception, Christ had to be a XX male, which meant he was sterile and likely lacked the desire to mate. Using words that he spoke in Matthew 19:12 NASB, Christ was a eunuch from his mother's womb.

(3) In John 1:1-4, 14 NASB, the apostle John talks about Christ being the Word of God in terms that many people have understood in a mystical, magical way. They believe that somehow Christ, as a person, was present when God created matter in the Universe. Let's look at what John wrote:

> *In the beginning was the Word, and the Word was with God, and the Word was God. He was in the beginning with God. All things came into being through Him, and apart from Him nothing came into being*

> *that has come into being. In Him was life, and the life was the Light of men. ... And the Word became flesh, and dwelt among us ...*

The exact process that started and still maintains the formation of matter in the Universe also facilitated the conception of Christ. Because John didn't understand who God is and what God generates, he saw through the glass darkly.[6] Thus, he couldn't know that the vibrations (words) that created matter were the same vibrations that created Christ. John didn't know that Christ was a new creation—in scientific terms, a new human species because of the changes in his DNA—that involved both a human being (Mary) and God.

(4) Christ acknowledged that he was different from other human beings as he called himself the Son of Man and the Son of God throughout his ministry. One of the main thrusts of what he preached was that God is a Father and doesn't need appeasement. An example was the Sermon on the Mount (Matthew 6 NASB) when Christ spoke of the relationship of father and child instead of deity and worshiper. Although some Old Testament writers acknowledged God as a Father, having Christ talk about God's fatherhood was unusual. The concept diverged from the teachings of the religious leaders and caused conflict. The notion that a person could speak directly to God alone in a prayer closet[7] took faith from being a public exercise into a secret relationship. As his ministry expanded, Christ continued emphasizing the love relationship between God and the human race—he spoke of God's love for the world[8] and challenged his listeners to love God with everything they had.[9]

(5) The second main thrust by Christ came with the concept that the measure of faith was how that person treated others and not the person's religious practices. After he asserted that loving God with everything one has was the greatest commandment, Christ then said the second commandment was similar: "*You shall love your neighbor as*

yourself."[10] Earlier in his ministry, he had already told his listeners not to judge others because the same measure would judge them.[11] Christ touched on this wisdom in several of his parables, particularly the story of the rich man and Lazarus.[12] Christ spoke of a rich man who had ignored a beggar named Lazarus who had been laid at his gate. After they both died, Lazarus was comforted by Father Abraham, and the rich man was in agony. Abraham refused to do what the rich man had refused to do with Lazarus—Abraham refused to bring the rich man comfort as the rich man had refused to bring Lazarus comfort. The rich man had judged Lazarus unworthy of being comforted, so Abraham judged the rich man similarly. Christ's point was that what went around came around. [If we love our neighbors, we won't judge them because love doesn't demand its way. If we give love, love will come back to us.]

(6) As far as the religious practices went, Christ challenged the tendency of the religious leaders to rigidly keep the rules (traditions) at the expense of meeting the needs of the people. Christ once zeroed in on excusing lack of care for parents with the pretext of giving the money to God.[13] Christ didn't hold back with his condemnation of this fraud, "*... And by this you invalidated the word of God for the sake of your tradition. You hypocrites ...*" .[14] Later, Christ tore into the religious leaders in Jerusalem and described several other ways that those leaders demonstrated hypocrisy.[15] He included the charge of them cleaning the outside of their walks of faith while ignoring the filthiness within themselves. Christ admonished them in no uncertain terms: hypocrites, fools, blind guides, lawless, serpents, vipers, and murderers. Christ wasn't raging against individuals, per se, but against the system that drew individuals into hypocrisy, that put rules ahead of people. As Christ said in an earlier situation, "*The Sabbath was made for man, and not man for the Sabbath.*"[16] Needless to say, the religious leaders were angry with Christ's accusations.

(7) During his ministry, Christ did more than teach. The four Gospels describe the many "miracles" he accomplished. While some people look at those miracles as almost magical happenings, physical realities were the basis of the good things that happened. The writer of the Gospel of Luke gave a hint of those realities as he described one instance in Luke 6:17-19 NASB when a great crowd of people came to where Christ was for the healing of their diseases and curing of being troubled by unclean spirits. The writer related, *"And all the people were trying to touch Him, for power was coming from Him and healing them all."* Then, in Mark 5:25-34 NASB, an ailing woman touched Christ's cloak and was healed. Christ immediately knew what had happened because he had perceived that power had gone out of him. That power went back to his conception—when the Holy Spirit overshadowed Mary and Christ came into being, everything that God generates infused the beginning cells of Christ's physical body. As he developed in the womb, was born, and matured, God's energy, life, love, vibrations, and good saturated Christ's being. That saturated power is what brought about the miracles that benefited many people.

(8) A few of Christ's miracles may not have happened because energy flowed out of him. Instead, he used wisdom to accomplish what his disciples thought was undoable. In Mark 6:33-44 and John 6:1-14, the writers talk about Christ ministering to a crowd of about five thousand people. As the hour grew late, he challenged his disciples to find food for all the people. When the disciples responded that there was no way they could do so, Christ had them seat everyone in groups of fifty to a hundred people. Then, he took the five loaves and two fish that a young boy had offered, blessed the food, and began breaking the food into pieces to hand out. Long story short, everyone ate until they were satisfied, and there were twelve baskets of leftovers. Some people understand this incident as a miracle where Christ handled every piece of food, either eaten or gathered in the baskets. We need to stop and think—for Christ to have touched all the food, the process

would have taken a tremendous amount of time, especially with the disciples taking the food to various groups. How much more likely was it that, after Christ had the crowd seated in smaller groups and after he blessed the boy's lunch, the people in the groups decided to share the food that they had brought with them? If one young boy had brought a lunch with him, the chances are good that many other people also had brought food with them. With the crowd seated in smaller groups, there were more interactions between individuals and more opportunities to share. As it turned out, there was more than enough food to satisfy everyone's hunger.

(9) Although the energy of God infused Christ, the context of the era he was living in limited his understanding as a human being. Although he was innately aware that he was God's Son, he didn't know everything. Hebrews 5:8 NASB says that Christ learned obedience from the things he suffered. (The Greek word translated as "suffered" means to undergo, to experience.[17]) Christ learned from the things he experienced—we could say he had to learn to be the Son of God. That process began when he was a child, as evidenced by the anecdote in Luke 2:41-51 KJV. At the age of twelve, Christ spent several days in the temple in Jerusalem and interacted with the religious teachers. Ironically, he did not go on to become a religious leader but followed Joseph's footsteps and became a carpenter. So, Christ experienced everyday life as a regular person rather than as a religious leader. Before his ministry, he took his place as God's Son due to his interactions with God and not due to any religious practice. As we can gather from his teachings, Christ believed that religious practices interfered with people finding their places as sons of God.

(10) As he taught and ministered to his fellow human beings, Christ faced a problem: while he was infused with all that God generates (especially God's life), Christ couldn't infuse other people with that life. The most he could do was to heal their infirmities and diseases because he couldn't change people in their inner beings, in their hearts. Also,

since Christ was infused with God's life, he was immortal as God is immortal—John 5:26 NASB puts it this way, "*For just as the Father has life in Himself, even so He gave to the Son to also have life in Himself.*" If Christ had remained as he was when he was born, he would have never died while he watched all of his fellow human beings dying year after year after year. God and Christ had the answer to this problem, a solution that would involve Christ emptying himself of all that the Holy Spirit had infused him with at his conception. So, from early on in his ministry, Christ talked of this emptying that his death would manifest. During his last meal with his disciples,[18] Christ spoke the following words: "*... But I tell you the truth, it is to your advantage that I go away; for if I do not go away, the Helper will not come to you; but if I go, I will send Him to you ...*" (John 16:7 NASB, the Helper is the Holy Spirit). Christ knew that the only way his fellow human beings could participate in the life that he had within himself was to empty himself of that life. His death would be the next step in making God's life attainable for the human race.

Chapter Notes

[1] Luke 1-2 KJV
[2] Matthew 1 KJV
[3] Luke 1:30-34 NASB
[4] Luke 1:35 NASB
[5] Genesis 1 NASB
[6] I Corinthians 13:12 KJV
[7] Matthew 6:6 NASB
[8] John 3:16 NASB
[9] Matthew 22:37 NASB
[10] Matthew 22:39 NASB
[11] Matthew 7:1-2 NASB
[12] Luke 16:19-30 NASB

[13] Matthew 15:1-9 NASB
[14] Matthew 15:6-7 NASB
[15] Matthew 23:13-36 NASB
[16] Mark 2:27 NASB
[17] *πάσχω - Wiktionary*. (2021, March 19). Wiktionary.Org. https://en.wiktionary.org/wiki/%CF%80%CE%AC%CF%83%CF%87%CF%89
[18] John 13-17 NASB

Remember

None

Mini-Essay

Long before Science opened the heights and depths of Creation to us, some religious leaders settled on the concept of the Trinity as the foundation of the Christian religion. This Godhead consisted of the Father, the Son, and the Holy Spirit. This theory was the best way those leaders could express how God, Christ, and the Holy Spirit interacted. The leaders had no concept of God as the ultimate Generator, so they tried to explain reality mystically. [Now that we are learning about the physical realities of the Universe, visualizing God as an actual physical being is easier for us (think of Him as being infinitely more powerful than the brightest star).] The doctrine of the Trinity claims that Christ was/is God, but two realities refute that concept. First, if Christ were God, he would generate God's energy, the Holy Spirit. That would mean that Christ generated the Holy Spirit before his conception so that Mary could conceive him. How convoluted that scenario would have been. Second, there is no possible way a human body could contain all that God is, and Christ was a human being.

Likewise, the difference between God and the Holy Spirit is that God is the Generator and the Holy Spirit is what God produces. The Holy Spirit can't generate himself. Maybe it's time to replace the concept of the Trinity with a more probable reality: God is God, the Holy Spirit is God's energy, and Christ is the Son of God, whom Mary conceived when the Holy Spirit overshadowed her.

12

THE POWER OF THE RESURRECTION

Key Words: Christ, Death, Resurrection, God, Infused, Circle, Competition, Religious Leaders, Cross, Forgiveness, New Testament, Empty, Energy, Nothingness, Deception, Living Water, Somethingness, Life, Crucified, Holy Spirit, Mary, DNA, Regenerate, Three Days, Three Nights, Mary Magdalene, Ascension, Pay-off, Eternity, Heaven, Hell, Immortality, Enemy, String Theory, Sabbath, Day of Preparation, Aerodynamics, Elijah, Prophet, Martyrs, Altar, Valley, Dry Bones, Choice

The Model

(1) Christ needed to die a spectacular death to achieve his resurrection. Let's think about this. Being infused with all that God generates, Christ couldn't get sick and die, nor could he meet with an accident and die (God's energy protected him). Also, right before his ministry started, he had already turned down the opportunity to commit suicide when Lucifer suggested it.[1] The only way Christ could die was for him to choose to be killed. So, to create the opportunity for that to happen, Christ entered into a vicious circle of competition with the Jewish religious leaders. [That he healed people and worked miracles was enough

to start the competition.] Christ stoked contention by disparaging the leaders[a] and driving the money-changers out of the temple[2] (once with a scourge of cords[3]). Once Christ raised Lazarus from the dead,[4] the Jewish leaders knew they had to do something because they were afraid their whole religion would be at risk if Christ continued with his ministry.[5] So, the leaders did what God and Christ wanted them to do—they conspired together and then succeeded in their quest to kill Christ. As it turned out, that process included name-calling[6] and even a scourging,[7] both part of the payback of Christ's engagement in the vicious circle with the religious leaders. However, on the cross, Christ broke that vicious circle by forgiving all that participated in the killing because they didn't know (understand) what was going on.[8] [Forgiveness is one of the weapons that break the power of vicious circles within the human race.]

(2) The New Testament writers knew a little about the concept of Christ emptying himself of his "deity." However, they linked that emptying to Christ's conception with the thought that Christ had previously existed in the form of God and was made in the likeness of human beings.[9] However, Christ was a new creation and had not existed before his conception; he was God's Son, not His twin. Christ was created the same way the matter in the Universe was. Yet, Christ did empty himself as the writers believed, only he did it on the cross. The last statements Christ made before he died indicated the emptying process. Let's look at what his words meant:

- *"Father, into Your hands I commit my spirit."*[10] At this point, Christ chose death—by giving up the energy that enabled him to walk as the Son of God, Christ emptied himself to become strictly human. That emptying was the only way Christ could die.
- *"My God, my God, why have You forsaken me?"*[11] As the energy drained out of Christ's body, Lucifer's nothingness/deception swept in to finish the job of death. Christ was the one who

changed, not God, and he became the deception that God had forsaken the human race after Adam and Eve chose to eat the fruit. This deception had kept humans in bondage to the belief that they needed to appease God to convince Him to discontinue forsaking the world.

- *"I am thirsty."*[12] When Christ spoke those words, the Roman soldiers gave him sour wine to drink. The fact that he was thirsty was ironic because, earlier in his ministry, Christ spoke to a Samaritan woman about his ability to give people the well of living water, which would spring up to eternal life.[13] Christ's words as he died on the cross were an indication that the well of living water that flowed within himself was drying up. And, when a soldier pierced Christ's side after he died, both water and blood came out,[14] which represented the draining of the living water and physical life out of his body.
- *"It is finished."*[15] After Christ sipped the sour wine, he died and completed emptying himself of all that God had infused him with at his conception. But that wasn't all that happened. At that point, the Somethingness of God gave way to nothingness.

The whole point of the cross was Christ choosing to empty himself and, in the process, becoming the very deception that brought death to the human race in the first place. [If people believe that God has forsaken them and that they need to appease Him to have a good afterlife (Heaven), they won't understand the power of life.] Christ destroyed that deception and would soon prove the power of life.

(3) During his ministry, Christ talked about being crucified and then resurrected after three days and three nights. He even warned the religious leaders what would happen if they caused his death by using the metaphor of the temple to describe himself: *"Destroy this temple, and in three days I will raise it up."*[16] Notice that Christ didn't say God would raise him, but that he would raise himself. Let's put this together

with Acts 2:27 NASB, which describes God's intention towards Christ: "*Because You will not abandon my soul to Hades, nor allow Your Holy One to undergo decay.*" Even though Christ physically died, God had already provided the process needed to keep him from undergoing decomposition. Remember,[b] when the Holy Spirit overshadowed Mary, God melded two of her gametes together. At that point, He made changes in Christ's genome, changes that would later cause his DNA to regenerate his physical body upon death. Because Christ had consistently chosen to cleave to God until he was on the cross, his DNA never changed (unlike Adam and Eve's DNA, which gradually changed after they chose to let go of God). Because there had been no changes, the regeneration process quickly kicked in once Christ died, and his DNA never decayed. His regeneration took three days and three nights to complete. So, Christ woke up just a few hours shy of seventy-two hours from the time of his death.[*]

(4) According to the account found in John 20:11-17 NASB, Mary Magdalene was the first person to see Christ as she stood weeping outside the empty tomb. What's interesting about this encounter was that Christ told her to "*stop clinging to me*" (translated as "*touch me not*" in KJV) and gave a reason for Mary not to touch him—he had not yet ascended to his Father. This ascension may have been separate from the ascension his disciples later observed. Whether he ascended to a mountaintop or Infinity, there could have been a distinct purpose for Christ to be alone with God. While his DNA had regenerated his physical body, Christ was still empty of all that had drained from his body during the death process. The only way for him to be refilled was for him to be in God's presence where the Holy Spirit, the energy of God, could again infuse Christ's body with all that God generates. Once that infusion happened, Christ was wholly resurrected, never to die again. And we know that Christ became touchable because he appeared to a group of the disciples hours later and said to them, "*See my hands and my feet, that it is I myself; touch me and see, for a spirit does not have flesh*

and bones as you see that I have."[17] After that encounter, Christ appeared to the disciples numerous times over forty days before being lifted up as a cloud received him out of their sight.[18] Christ ascended to join God in Infinity, where he was seated at his Father's right hand.[19]

(5) Since Christ's death and resurrection took place, the focus of the interruption to his physical life has been on the death event. From the very beginning, some religious doctrines have emphasized that the only way we can come to God (i.e., the only way to appease God) is through the blood of Christ. Then, as believers kept dying, the pay-off for "accepting" Christ became spending Eternity in Heaven, while the consequence for "not accepting" Christ became spending Eternity in Hell. Christianity developed and became like many other religions: it centered on being the only way to God (competition, anyone?). Thus, because humans weren't finding immortality, the return of Christ almost became an afterthought, pushed well into the future. Christianity taught that God would eventually get so disgusted with the world that He would destroy the Earth, send everyone who believed in Christ to Heaven, and send everyone else to Hell. Additionally, the reality of people dying became so romanticized that many blamed God for those deaths: "God called them home. He needed them in Heaven ..." Some believers even began looking forward to when everyone dies and goes to Heaven. Some also look forward to death because the Earth isn't their home, and they want to be with Christ. This fixation with Christ's death as the focal point of God's plan and the death of believers as the ultimate goal of the Christian faith completely misses the whole point of the resurrection.

(6) The apostle Paul had a stern message for those believers who doubted the resurrection in 1 Corinthians 15 NASB (well worth a read) and made a strong case for the coming immortality for the human race. In verse 20, Paul likened death to being asleep and stated that Christ was the first fruit of those resurrected from death. Then, in verse 26, Paul left no doubt that death is an enemy to the human race with the

words, "*The last enemy that will be abolished is death.*" Paul's statement confirmed Christ's words in John 10:10 NASB, where he said the thief came to kill while he (Christ) came that we might have life. [Christ was talking about physical life, not some mystical, spiritual energy that happens in a far-off place called Heaven.] A statement from Proverbs warns about replacing God's focus on immortality with what seems right to us, "*There is a way which seems right to a man, but the end is the way of death.*"[20] Let's consider the concept of Heaven and Hell in the light of that statement. The only way a human could go to Heaven or go to Hell is via death. People who are alive are here on Earth and not in Heaven or Hell. Yes, it seems reasonable that good people (those who have "accepted" Christ) would go to Heaven and bad people (those who have "rejected" Christ) would go to Hell. But in this approach, the end of both choices is death. And death is contrary to the nature of God and the reality of the resurrected Christ.

(7) Here's a question we should consider: if Christ is alive, how can dead people be where he is? People who have died cannot interact with people who are living. Also, because God has given humans a choice between life and death, He won't override their choices—He won't override death.[**] (Let's think: the life God generates would automatically override human will if non-living humans were really with Christ, who is seated at God's right hand.) That would be contrary to God's nature. So, what happens when people die? Remember,[c] we previously discussed the string theory that scientists are working on and how those strings could manifest God's love. What if God attaches one of those strings to an individual's DNA at conception? Then, at death, God could take strands of the individuals' DNA (via the strings) to keep safely somewhere in the Universe or on Earth. No one would see it happen because both the strands of DNA and the strings would be too small to be perceived. [At the point of death, an individual's DNA is deactivated in their physical body but remains intact in the safe place until the resurrection of the dead takes place.]

(8) While the writers of the Bible talked about the resurrection of the dead, in 1 Corinthians 15:51-52 NASB, Paul went even further and wrote that there would be a group of people who never die:

> *"Behold, I tell you a mystery; we will not all sleep, but we will all be changed, in a moment, in a twinkling of an eye, at the last trumpet; for the trumpet will sound, and the dead will be raised imperishable, and we will be changed."*

Christ had said something similar to Martha right before he raised Lazarus from the dead (John 11:25-26 NASB): *"I am the resurrection and the life; he who believes in me will live even if he dies, and everyone who lives and believes in me will never die."* We could say there will be people who put on immortality without ever dying. From Paul's description, how that will happen seems like the waving of a magic wand—one moment those people are mortal, then, boom, they become immortal. We could say the same about Christ; one moment he was dead, and then he was alive. But, underneath the surface of the resurrection, we can see the process that regenerated Christ's physical body. Putting on immortality by the living and the dead may also be a process. We'll discuss the "how-it-will-happen" for those mechanisms in Chapter 14, but first, let's look at the Book of Job. Some biblical scholars believe it is the oldest book in the Bible, which is ironic because it lays out the whole story of God, humans, and religion as a five-act drama. Hint: there is a happy ending.

Chapter Notes

[1] Matthew 3:5-7 NASB
[2] Matthew 21:12-13 NASB
[3] John 2:11-17 NASB

[4] John 11 NASB
[5] John 11:47-53 NASB
[6] Matthew 26:67-68, Matthew 27:27-31 NASB
[7] Mark 15:15 NASB
[8] Luke 23:17 NASB
[9] Philippians 2:7 NASB
[10] Luke 23:46 NASB
[11] Mark 15:34 NASB
[12] John 19:28 NASB
[13] John 4:13-14 NASB
[14] John 19:34 NASB
[15] John 19:30 NASB
[16] John 2:19 NASB
[17] Luke 24:39 NASB
[18] Acts 1:9 NASB
[19] Hebrews 12:2 NASB
[20] Proverbs 14:12, 16:25 NASB

* The Sabbath ran from Friday sunset to Saturday sunset, while the day of preparation ran from Thursday sunset to Friday sunset. Christ was crucified just before the day of preparation began. Mark 15:42 NASB relates the specific time when Joseph of Arimathea came to retrieve Christ's body: "*When the evening had already come, because it was the preparation day, that is the day before the Sabbath.*" Christ had repeatedly said he would be dead for three days and three nights: Thursday late morning/afternoon, Thursday night, Friday, Friday night, Saturday, and Saturday night. He came alive early Sunday morning.

** We'll talk about what happens at the Great White Throne Judgement in Chapter 15. Even at that point when death is swallowed up forever, God leaves the decision up to the individual: immortality or non-existence.

Remember

[a] Chapter 11, Paragraph 6
[b] Chapter 11, Paragraph 2
[c] Chapter 2, Paragraph 6

Mini-Essay

The aerodynamics of Christ's ascension would be interesting to understand—a living human being went through the Earth's atmosphere and the Universe into Infinity/Eternity. The possibility of a mechanical ascent is legitimate because the Old Testament hints at such vehicles. In 2 Kings 2:11 NASB, when Elijah was taken up to God, he departed from Elisha in a chariot of fire and horses of fire (perhaps some sort of a space vehicle) and then went up in a whirlwind (perhaps a launch). Then, in Ezekiel 1, the prophet described a vision with wind, fire, bright light, glowing metal, loud sounds, and figures with faces and wings that moved but never turned. The possibility that the four figures were space vehicles seems reasonable. So, Christ could have ascended mechanically—the biblical accounts of his ascension (Mark 16:19, Luke 24:50-51, Acts 1:9 NASB) state that he was taken up, but those accounts don't indicate how that happened.

If the concept of people going to Heaven after death were reality, the ones most deserving of the honor would undoubtedly be those people who have been killed for their faith in Christ. However, Revelation 6:9-11 KJV indicates that the martyrs aren't anywhere near God and Christ:

> *And when he had opened the fifth seal, I saw under the altar the souls of them that were slain for the Word of God, and for the testimony which they held: And they cried with a loud voice, saying,*

"How long, O Lord, holy and true, dost thou not judge and avenge our blood on them that dwell on the earth?"

And, white robes were given unto every one of them; and it was said unto them, that they should rest yet for a little season, until their fellowservants also and their brethren, that should be killed as they were, should be fulfilled.

The martyrs' location under the altar would indicate their deactivated DNA (souls) are either somewhere in the Universe or somewhere on Earth. Added to this, the prophet Ezekiel wrote of a vision God had given him: in this vision—recorded in Ezekiel 37:1-14 NASB—he saw a valley full of dry bones, over which God told him to prophesy. The prophet obeyed God, and the bones came together, fleshed out, and came to life. Those dry bones (dead people) had turned into a great army of people. Listen to what God told Ezekiel to prophesy to the whole house of Israel (the placeholder for the human race) in verses 12-14:

Therefore prophesy and say unto them, "Thus says the Lord God, 'Behold, I will open your graves and cause you to come up out of your graves, My people; and I will bring you into the land of Israel. Then you will know that I am the Lord, when I have opened your graves and caused you to come out of your graves, My people. I will put My Spirit within you and you will come to life, and I will place you on your own land. Then you will know that I, the Lord, have spoken and done it," declares the Lord.

Christ extended God's promises to Israel to the whole human race. We are all God's people.

Death separates humans from God and everything He generates. When we die, we don't magically come into His presence. If we did, He would negate the choices we have made. Deuteronomy 30:15-20 KJV is clear-cut: the choice between life and death is ours to make. When Moses wrote the passage, the promise of cleaving to God was long life; after Christ's resurrection, the promise became immortality.

Suggested Reading

Hertzenberg, S. (2021, November 30). *What is the Oldest Book in the Bible?* https://www.beliefnet.com/faiths/christianity/what-is-the-oldest-book-in-the-bible.aspx

String Theory: The Mathematics of Moving Beyond Current Physics. (2019, March 2).
https://www.thoughtco.com/what-is-string-theory-2699363

13

JOB: PASSIVE FAITH BECOMES ACTIVE FAITH

Key Words: Job, Bible, Righteousness, God, Drama, Faithfulness, Lucifer, Deceptions, Destruction, Friends, Power, Elihu, Wicked, Afflictions, Instruction, Whirlwind, Behemoth, Leviathan, Sword, Fear, Pride, Universe, Purpose, Christ, Cause, Unrighteous, Good, Resurrection, Nothingness, Promise, Immortality

The Model

(1) The Book of Job could claim a prize for being one of the most misunderstood books of the Bible. According to some people, this story of a righteous man, devastated by terrible loss and physical affliction, means that the righteous must bear suffering because God allows it to happen. The tendency to read the Bible line for line and assume that each line is biblical truth can cause us to miss the whole point of Job's experience. So, to understand the depths of Job's suffering and what he learned, we're going to look at the book as a five-act drama. We'll allow for the feelings and confusion that Job's destruction brought to him and those around him.

(2) *Act 1, chapters 1-2:* Job was, by all accounts, a righteous man who made appeasing God and doing good his top priorities. For all that Job did as a righteous person, one thing he didn't have was confidence in God, as evidenced by his need to appease Him with offerings for any possible wrong-doings by his children. When his friends first showed up to comfort him, Job said, "*For the thing which I greatly feared is come upon me, and that which I was afraid of is come unto me. I was not in safety, neither had I rest, neither was I quiet; yet, trouble came.*"[1] Job's attitude was that he had to continually prove himself to God with the fear that God may turn His back on him or worse. For some reason, Job didn't understand God's faithfulness (nor was he expecting it), and Lucifer (aka Satan) picked up on Job's lack of understanding and was intent on going after the man. [Although the wording in Job 1:8 KJV does sound like God may have deliberately set Lucifer's attention on Job, the literal Hebrew (per the KJV note) reads God's question to Lucifer as "*Hast thou set thy heart on ...*".] However, God was able to set some limits on Lucifer because of Job's actions, especially concerning how he treated those in need.[2,3] Also, remember,[a] Lucifer's words are always deceptions—in his exchange with God, he accused God of putting a hedge around Job and told God to destroy all that the man had. The reality? God doesn't build barriers around people because that would infringe on their rights to choose; God also never destroys anything. In His response to Lucifer's words, God spoke the reality: everything Job has was in Lucifer's hands. We need to realize that nothingness can come against us at any time without our permission. God will not stop the destruction, but He will help us and give us all we need to avoid or end the destruction ourselves if we seek His help. Job didn't understand this, so his response to both of the devastating onslaughts by Lucifer was to blame God. Sure, he couched his response as faith in God despite everything that happened, but the reality was Job blamed God for everything. We'll see this magnified in Act Two.

(3) *Act 2, Chapters 3-31:* As Job processed the total devastation of his life, three of his friends—Eliphaz, Bildad, and Zophar—came to him to sympathize with him and to bring comfort. Job and his three friends agreed that God had been the one to inflict the destruction upon him. What they differed on was whether Job deserved the devastation or not. All three "comforters" insisted that Job was the one in the wrong, with Eliphaz (the first one to speak up) stating, "... *Remember now, whoever perished being innocent? Or where were the upright destroyed? According to what I have seen, those who plow iniquity and those who sow trouble harvest ...".*[4] As the argument grew with his friends, Job's responses centered on his belief that God had acted unjustly and that he (Job) had done nothing wrong. For instance, in Job 9:17-24 NASB, the man said, "*For He bruises me with a tempest and multiplies my wound without cause. He will not allow me to get my breath, but saturates me with bitterness. ... Though I am guiltless, He will declare me guilty. ... If it is not He, then who is it?*" At this point, the common thread behind religious practices was that God needed to be appeased and persuaded to bless people. When destruction came, the religious folks believed that the victim had done some great wrong or had not sufficiently mollified God. They thought that God held all the power over everything, so Job had no other alternative but to blame God for everything. And, because the devastation was so complete and his friends' condemnation so fierce, Job was all over the place about how he should react to God and the destruction. In Job 13:1-4, 15 NASB, his anger became directed at his friends and God:

> *Behold, my eye has seen all this* (God's power), *my ear has heard and understood it. What you know I also know; I am not inferior to you. But, I would speak to the Almighty, and I desire to argue with God. But you smear with lies; you are all worthless physicians. ... Though He slay me, I will hope in Him. Nevertheless I will argue my ways before Him.*"

Job reached a point where he didn't want to continue arguing with his friends; instead, he desired to contend with God. However, his friends continued to pressure Job until he had had enough. He listed the good he had done in the past and the evil he had not done.[5] That list was enough to quiet his friends.

(4) *Act 3, Chapters 32-37:* After listening carefully to the conversation between Job and his friends, Elihu—whose name means "God Himself"—was angry with everyone involved. Job had justified himself over God, and his three friends had condemned Job without providing any answers to Job's questions. Elihu had quite a lot to say, so let's look at some highlights of his monologue:

- In response to Job's friends, Elihu pointed out that their words had been in vain because none of them had answered Job's words. He said he wouldn't use their arguments to respond to Job's words.[6] Before explaining concepts to Job, Elihu challenged the older man to refute him. After Elihu emphasized his humanity,[7] he summarized Job's words, per Job 33:8-10 NASB: *"Surely you have spoken in my hearing, and I have heard the sound of your words: 'I am pure, without transgression; I am innocent and there is no guilt in me. Behold, He invents pretexts against me; He counts me as His enemy.'"* That, in a nutshell, is what Job believed.
- Elihu goes on in Job 34:10-13 NASB to state categorically:

> *"Far be it from God to do wickedness, and from the Almighty to do wrong. For He pays a man according to his work, and makes him find it according to his way. Surely God will not act wickedly, and the Almighty pervert justice. Who gave Him authority over the earth? And who has laid on Him the whole world?"*

God honors our choices (our words and actions, aka our works). He won't override those choices.

- However, God will reach out to those who have made mixed choices—on the one hand, choosing God and, on the other hand, choosing nothingness. Elihu spoke on this important concept in Job 36:7-12 NASB:

> *"He does not withdraw His eyes from the righteous; but with kings on the throne (,) He has seated them forever, and they are exalted. And if they are bound in fetters, and are caught in the cords of affliction, then He declares to them their work and their transgressions, that they have magnified themselves.*
>
> *"He opens their ear to instruction, and commands that they return from evil. If they hear and serve Him, they will end their days in prosperity and their years in pleasures. But if they do not hear, they shall perish by the sword and they will die without knowledge."*

God speaks, but we have the choice of whether we hear Him or not. He will work with us if we listen.

Surprisingly enough, neither Job nor his friends argued with anything Elihu said.

(5) *Act 4, Job 38-41*: God spoke to Job out of a whirlwind—which included energy (wind) and vibrations (words)—and told him to stand up and answer His questions. Then, God described His abilities in the form of questions to Job and mentioned several things that had come up in the discussions between Job and his friends and in Elihu's speech. But, God went further and expounded on the extent of His abilities. [Why did He do that? Job's life and faith had undergone a severe battering, so God rebuilt Job's faith with the reminder that God is God.] About

halfway through what He was speaking, God paused and asked Job if he would contend with Him. Job answered that he had already spoken and wouldn't say anything more. God's answer to Job was, "*Will you really annul My judgment? Will you condemn Me that you might be justified?*"[8] Next, God got to the crux of the destruction that had engulfed Job—He spoke of the behemoth and the leviathan (metaphors for Lucifer):

- Job 40:19 NASB: "*He (the behemoth) is first of the ways of God; let his maker bring near his sword.*" God is Lucifer's maker, and the sword is God's word, His vibrations. [In Hebrews 4:12 NASB, God's word is compared to a two-edged sword while, in Revelation 1:16 NASB, a sharp two-edged sword came out of Christ's mouth.] God's sword, His word (vibrations), is the only effective weapon against Lucifer.
- Job 41:24-26 NASB: "*His (the leviathan's) heart is as hard as a stone, even as hard as a lower millstone. When he raises himself up, the mighty fear; because of the crashing, they are bewildered. The sword that reaches him cannot avail, nor the spear, the dart or the javelin.*" Lucifer's heart is hard against humans, and there is no human weapon that will stop him.
- Job 41:33-34 NASB: "*Nothing on the earth is like him, one made without fear. He looks on everything that is high; he is king over all the sons of pride.*" Lucifer is not afraid of anything humans can do to him, nor is he afraid of God. God doesn't and won't occupy the Universe, so Lucifer thinks his kingdom is safe. Also, Lucifer reigns over all those who are proud (who rely on their actions/goodness). So, on its own, the human race has no defense against Lucifer.

For the first time since Job and his three friends started their interaction, the concept of "someone" other than God causing destruction, death, and loss entered the discussion.

(6) *Act 5, Job 42:* Job's first words to God after He stopped speaking were, *"I know that You can do all things, and that no purpose of Yours can be thwarted."*[9] What a revelation—God generates somethingness while Lucifer demolishes things into nothingness. God's purpose is that humans have the freedom to choose between life and death, between blessings and cursings. His purpose is that He won't force people to know and understand Him. [God was able to speak to Job because he called God out; the man had gone from "God gives and takes away" to "I want to argue with You, God!"] Next, Job admitted that his understanding of God had been all wrong, and he spoke the words that God wants to hear from every human being: *"Hear, now, and I will speak; I will ask You, and You instruct me. I have heard of You by the hearing of the ear; but now my eye sees You; therefore I retract, and I repent in dust and ashes."*[10] Job finally understood that his religious practices, which came from what other people had said, could never measure up to the one-on-one relationship with God. [The difference is similar to hearing a description of a banana split versus eating a banana split. There is no comparison between those two experiences.] Job's life immediately changed as God gave him instructions on how to mend his relationship with his three friends.[11] And, Job kept interacting with God, so he never faced destruction again in his lifetime.[12]

(7) As the probable first book of the Bible ever written, Job may symbolize the human race. Lucifer (nothingness) doesn't need a reason to steal, kill and destroy[13] while God is ready to help us individually to find answers if we turn to Him.[14] The biblical history of the Hebrews and the early Christian church shows the ups and downs of this dichotomy that persists in our world. And the hope that our religious practices will appease God has continued since then, even though Christ is resurrected and has made the way for immortality for the human race (which Job didn't experience). How do we get to where God wants us to be?

Chapter Notes

[1] Job 3:25-26 KJV
[2] Job 29:12-17 KJV
[3] Psalm 41:1 KJV
[4] Job 4 NASB, especially verses 7-8
[5] Job 29, 31 NASB
[6] Job 32:11-13 NASB
[7] Job 33:5-7 NASB
[8] Job 40:8 NASB
[9] Job 42:1 NASB
[10] Job 42:4-6 NASB
[11] Job 42:7-10 NASB
[12] Job 42:12-17 NASB
[13] John 10:10 NASB
[14] Matthew 7:7-8 KJV

Remember

[a] Chapter 7, Paragraph 2

Mini-Essay

Some people connect Job with the suffering of the righteous. They look at righteousness from the human context—any person who keeps the rules and does good qualifies as righteous. Yet, righteousness doesn't originate with what people do; instead, it has to do with who they are. Listen to Matthew 6:33 KJV: *"But seek ye first the kingdom of God and His righteousness; and all these things shall be added to you."* [God is and generates righteousness.] As discussed, Job was more concerned with maintaining his self-righteousness than God's. When he accused

God of causing the destruction that came into his life, Job slammed God for being unrighteous, which allowed God to confront him. Job's self-righteousness crumbled during that confrontation as God's righteousness soared and turned things to Job's good. [Interacting directly with God works to our advantage.]

Before Christ came into being, there was no way for people to become God's righteousness. Psalm 14:2-3 KJV relates,

> *"The Lord looked down from heaven upon the children of men, to see if there were any that did understand, and seek God. They are all gone aside, they are all together become filthy; there is none that doeth good, no, not one."*

As we see with Job, his focus wasn't on seeking and understanding God. Job was centered on himself and all the good that he was doing. Because of this focus, the good that Job had done disintegrated when devastation hit. Fortunately for us, Christ changed everything. When we take advantage of all that he accomplished through his resurrection, we have the opportunity to become the righteousness of God instead of relying on our righteousness. Here's what 2 Corinthians 5:21 KJV says: *"For he (God) hath made (Christ) to be sin for us, who knew no sin; that we might be made the righteousness of God in him (Christ)."* When he died, Christ became the nothingness that Lucifer's deceptions pushed the human race into, even though he (Christ) had never partaken in that nothingness. Because Christ chose to die with God's promise that he would regain his life, we have the opportunity to be made the righteousness of God.

Why is this important? Proverbs 12:28 NASB states the benefits of God's righteousness: *"In the way of righteousness is life, and in its pathway there is no death."* No death equals immortality. Becoming God's righteousness is how we attain immortality. Christ's resurrection is the lynchpin of immortality as he defeated death and made way for us to become the righteousness of God. Becoming (or being made) denotes a process, so our next topic is the way we become the righteousness of God.

14

THE NEW OPTION

Key Words: Christ, Enemies, Death, Deception, Lucifer, Adam, Eve, God, Resurrection, Immortality, Nicodemus, Born Again, Pentecost, Holy Spirit, Energy, Frequency, Vibration, Sensory Receptors, Sound Waves, Speaking in Tongues, Paul, Knowledge, Understanding, Wisdom, Heaven, Prayer Language, Prayer Closet, Rituals, Reason, Judgments, Universe, Earth, Life, Logic, Destructive, Words, Actions, Reality

The Model

(1) In 1 Corinthians 15:25-26 NASB, the Apostle Paul talked about the defeat of the enemies of Christ (deception, loss, hate, destruction, etc.) with death as the last one to be conquered. Paul's words tell us that there are no magic wands involved with immortality; the manifestation of endless life is a process. We've already discussed[a] how the deceptions of Lucifer brought death into the Earth, as he succeeded in manipulating Adam and Eve into letting go of God. At that point, death began taking hold of the world and then ruled over the world until Christ overcame death through the resurrection of his physical body. As Paul wrote in 1 Corinthians 15:20-22 NASB:

> "*But now Christ has been raised from the dead, the first fruits of those who are asleep. For since by a man came death, by a man also came the resurrection of the dead. For as in Adam all die, so also in Christ all will be made alive.*"

Remember,[b] Paul indicated that the putting on of immortality would happen in two different ways: those who are living at that time will undergo a change in their physical bodies, and those who are dead will be resurrected. So, two different processes will happen, and both will involve a "cleansing" system that will winnow out every bit of deception, loss, hate, destruction, etc., that has plagued the human race since the Garden of Eden. As Matthew 3:11-12 NASB says:

> "*... He (Christ) will baptize you with the Holy Spirit and fire. His winnowing fork is in His hand, and He will thoroughly clear His threshing floor; and He will gather His wheat into the barn, but He will burn up the chaff (the nothingness) with unquenchable fire.*"

Through this mechanical cleansing provided by all that Christ accomplished through his resurrection, God will clear out everything in our lives that would stop immortality from being ours.

(2) To understand how the process of putting on immortality starts, we need to look at the beginning of a night-time conversation that Christ had with Nicodemus, a Pharisee, as recorded in John 3.1-8 NASB. Christ's first words were, "*Truly, truly, I say to you, unless one is born again he cannot see the kingdom of God.*" [What Christ meant was that human beings were unable to see (as in perceive) the kingdom of God. Everything God generated remained on the outside of people and generally made no difference to who they were on the inside.] After Nicodemus questioned the process of being born a second time from a mother's womb, Christ explained:

"Truly, truly, I say unto you, unless one is born of water (the physical birth) and of the Spirit (the rebirth) he cannot enter into the kingdom of God. That which is born of the flesh is flesh, and that which is born of the Spirit is spirit. Do not be amazed that I said unto you, 'You must be born again. The wind blows where it wishes and you hear the sound of it, but do not know where it comes from and where it is going; so is everyone who is born of the Spirit."

In talking about the second birth, Christ likened the Holy Spirit to the wind, to energy. Let's look at what happened to Christ's followers in Jerusalem shortly after he had ascended to God (in Acts 2:1-4 NASB):

When the day of Pentecost had come, they were all together in one place. And suddenly there came from heaven a noise like a violent rushing wind, and it filled the whole house where they were sitting. And there appeared to them tongues as of fire distributing themselves, and they rested on each one of them. And they were all filled with the Holy Spirit and began to speak with other tongues, as the Spirit was giving them utterance.

Just as it was when the matter in the Universe came into being, so it was on that day in Jerusalem. Energy, frequency, and vibration (wind/ the Holy Spirit, fire/light, and words) were the forces that brought a new reality—the new birth—to that group of people. Remember,[c] those forces were also what brought Christ into being. All three events—Creation, the conception of Christ, and the Day of Pentecost—were all about God creating something new. The Apostle Paul recognized the newness that Pentecost brought as he wrote in 2 Corinthians 5:17 NASB: "*Therefore if anyone is in Christ, he is a new creature; the old things passed away; behold new things have come.*" [Remember,[d] God spoke through Ezekiel that He would give us new hearts and spirits[1]—this event was the beginning of the manifestation of that promise.]

(3) So, Pentecost changed people—the Holy Spirit ushered in a new option available to individuals by creating a system of sensory receptors tuned to all that God generates. [Lest we think that's not reality, scientists have already determined that the human body has sensory receptors that respond to electromagnetic radiation, electric fields, infrared radiation, magnetic fields, visible light, and ultraviolet light, among other things. With about 100 billion neurons in an individual, the addition of new sensory receptors in a person may not be physically noticed, either by the person or the people around them.] These new sensory receptors give a person the same ability to perceive what God generates as Christ had when he was born. These sensory receptors allow a person to become a "son" of God. However, a massive difference between Christ and the rest of us human beings still exists even with the new birth. Remember,[e] when the Holy Spirit overshadowed Mary and two of her gametes melded together, every bit of nothingness in Christ's DNA was swallowed up by God's generated power. His DNA became a clean slate (like Adam's DNA), and the Holy Spirit created the new sensory receptor system that enabled Christ to perceive God. The rest of us are physically born with a legacy from our DNA and the inherent lack of a sensory receptor system tuned into God. As we mature, deceptions enter the picture and manipulate our words and actions. While Christ could always understand reality (until right before he died), everyone else was and is blind to it.

(4) God needs to communicate effectively with those who choose to be born again. So, the new sensory receptor systems (in people who are born again) include elements that catch some of the vibrations that He generates. As they travel through the sensory system, His vibrations become sound waves coming out of the believers' mouths.[2] On the Day of Pentecost, this new phenomenon was labeled as "speaking in tongues" because the words weren't learned from any other human being but were given to individuals by God. At first, the new believers publicly used their new abilities in meetings, which caused chaos to

such an extent that the Apostle Paul addressed the issue. He wrote to the Corinthians: "*I thank God, I speak in tongues more than you all; however, in church I desire to speak five words with my mind so that I may instruct others also, rather than ten thousand words in a tongue.*"[3] The new believers didn't understand the mechanics and purpose behind speaking in tongues—they thought it was a sign/gift of the Holy Spirit and didn't connect the languages to God's knowledge, understanding, and wisdom. This lack of comprehension allowed Lucifer to pull believers away from the communication system God had set up for them. Before long, religious practices began taking the place of one-on-one communication between each believer and God. What people learned about God and Christ came from others instead of from God to each individual. This change of focus caused believers to shift from the hope of immortality to the hope of Heaven.

(5) From his comments in 1 Corinthians 14, we can infer that Paul spent a lot of time speaking tongues in private, where the communication system became a prayer language between him and God. Since Paul was responsible for much of the then-new (and incomplete) information about the reality that Christ's resurrection offered to the human race, the time Paul spent speaking in tongues seems to have paid off for the world. Unfortunately, despite God's desire for one-on-one relationships with individuals, many believers in Christ came to believe that God has chosen only certain people (like Paul) to receive information about the faith. These believers relied on what was written by others to guide their faith in God rather than communicating with God directly through the individual prayer languages put in place on Pentecost. However, God has never rescinded that communication system from the human race, nor has He limited who could utilize it. God isn't a respecter of people,[4] and He'll reward everyone who diligently seeks Him.[5] Nothing stops us from receiving knowledge, understanding, and wisdom from God through those prayer languages except our choices and lacks of effort.

(6) While delivering the Sermon on the Mount, Christ made a salient point about immortality: "*Enter through the narrow gate; for the gate is wide and the way is broad that leads to destruction, and there are many who enter through it. For the gate is small and the way is narrow that leads to life, and there are few who find it.*"[6] If we put this together with words he spoke earlier in his sermon, we can see what the narrow way is: "*But you, when you pray go into your inner room, close the door and pray to your Father who is in secret, and your Father who sees what is done in secret will reward you.*"[7] Let's remember what Deuteronomy 30:20 KJV says, "*That thou mayest love the Lord thy God, and that thou mayest obey His voice, and that thou mayest cleave unto Him: for He is thy life ...*". Life—immortality—is wrapped up in God. This reality is why our prayer closets and prayer languages are so important. They remove us from the vanity of rituals and the repetition of words written long ago by people who were limited in their understandings. Listen to what Christ said to his disciples shortly before he endured crucifixion in John 14:23-26 NASB:

> *"If anyone loves me, he will keep my word; and my Father will love him, and We will come to him and make Our abode with him. He who does not love me does not keep my words; and the word which you hear is not mine, but the Father's who sent me. These things I have spoken to you while abiding with you. But the Helper, the Holy Spirit, whom the Father will send in my name, He will teach you all things, and bring to your remembrance all that I said unto you."*

Christ spoke to the individual's interaction with God, which centers on words (vibrations). The Holy Spirit brings everything God generates into our lives; that makes Him the Helper. These words that Christ spoke connect with Psalm 91 KJV (well worth a read) and the promises given to the one who dwells in the secret place of the Most High, to the one who sets their love upon God.

(7) We've previously discussed[e] how God made it clear, through the prophet Isaiah, that He hated the religious practices of the Hebrews.[8] Again, God didn't just leave the Hebrews hanging—He went on to explain what He was looking for in Isaiah 1:18-20 NASB:

> *"Come now, and let us reason together," says the Lord, "Though your sins are scarlet, they will be white as snow; thought they are red like crimson, they will be like wool. If you consent and obey, you will eat the best of the land; but if you refuse and rebel, you will be devoured by the sword." Truly the mouth of the Lord has spoken.*

In the place of all the religious practices, God wants to reason with us (individually)—He wants to help us think, understand, and form judgments (of ourselves, not of other people) by reason. When we look at the Universe, the Earth, and the life forms on our planet, we can see logic everywhere in the systems that operate, sometimes together. God wants to apply that same logic to our lives. Then, we can be set free from destructive words and actions (what we call sin) and deal with everything that pushes us to separate from God and immortality. In this process of reasoning with God, we never lose our choice of heeding God's input or shunning it. If we choose to listen to God and depend on His guidance, our lives will show it. If we shun Him, our lives will show that choice.

(8) One of the hindrances to our reasoning with God is that the answers to our questions and problems are often at the calculus level while we are still learning our multiplication tables. God acknowledges this in Isaiah 28:9-13 NASB:

> *To whom would He teach knowledge, and to whom would He interpret the message? Those just weaned from milk? Those just taken from the breast?*

> *For He says, "Order on order, order on order, line on line, line on line, a little here, a little there." (KJV substitutes "precept" for "order.")*
>
> *Indeed, He will speak to this people through stammering lips and a foreign tongue, He who said to them, "Here is rest, give rest to the weary," and, "Here is repose," but they would not listen.*
>
> *So, the word of the Lord to them will be, "Order (KJV precept) on order, order on order, line on line, line on line, a little here, a little there," that they may go and stumble backward, be broken, snared and taken captive.*

If we are going to reason with God, we need to be aware that there will be times when He seems to be stuttering or speaking in a foreign language (not the prayer language) because we don't have the skills to understand what He is saying. In those situations, we have the choice of pressing into understanding what God is teaching us or giving up. If we choose to go on, we may have to return to and adjust some concepts that we've already accepted as reality. If we decide not to do that work, then God won't be able to bring us beyond the misconceptions we have. We'll stay in the mire of deception. Sometimes, we might need input from another person or other people—Job needed Elihu to help him understand. If we need human help (psychological, medical, financial, etc.), we can ask God to lead us to that help. At the same time, we should stay in our prayer closets and seek God with everything we have because no human can take the place of God.

(9) So, the resurrection of Christ opened up a new option for the human race. We're no longer unable to perceive God if we choose this experience of the new birth and the prayer language. With the creation of sensory receptors and the ability to directly speak God's vibrations aloud, we can come to the place of reasoning with Him, which may well lead us to immortality. We also need to understand that putting on immortality is like giving birth. Christ predicted that intense birth

pangs[11] would precede the manifestation of that everlasting life. These pangs will push the human race towards believing that all is lost.

Chapter Notes

[1] Ezekiel 36:26 NASB
[2] Acts 2:4 NASB
[3] I Corinthians 14:18-19 NASB
[4] Romans 2:11 KJV
[5] Hebrews 11:6 KJV
[6] Matthew 7:13-14 NASB
[7] Matthew 6:6 NASB
[8] Isaiah 1:10-15 NASB
[9] Matthew 24:8 NASB

Remember

[a] Chapter 7, Paragraphs 1-7
[b] Chapter 12, Paragraph 8
[c] Chapter 11, Paragraph 2
[d] Chapter 10, Paragraph 9
[e] Chapter 11, Paragraph 2
[f] Chapter 8, Mini-Essay

Mini-Essay

AN IMPORTANT LESSON: ***It's not what we do in front of others that counts; it's what we do in our prayer closets.***

Suggested Reading

Libretexts. (2020, August 14). *12.2A: Classification of Receptors by Stimulus.* https://med.libretexts.org/Bookshelves/Anatomy_and_Physiology/Book%3A_Anatomy_and_Physiology_(Boundless)/12%3A_Peripheral_Nervous_System/12.2%3A_Sensory_Receptors/12.2A%3A__Classification_of_Receptors_by_Stimulus

The nervous system: more than 90,000 miles of sensations! - Visual Dictionary. http://www.ikonet.com/en/visualdictionary/static/us/the_nervous_system

15

MYTH, REALITY, AND TRANSITION

Key Words: Christ, Tribulation, Rapture, God, Great Flood, Noah, Love, Myth, Eternal Hell, Fire, Infinity, Light, Covering, Deception, Human, Immortality, Resurrection, Great White Throne Judgment, Reality, Abomination of Desolation, Holy Place, Religious Practices, Death, Competitiveness, Beast, Sea, Earth, Mark, Number, Babylon, Scammers, Vicious Circles, Words, Actions, Karma, Confusion, King, Lord, Pilate, Inner Self, White Horse, Righteousness, New Name, Seven Churches, Individual

The Model

(1) In Matthew 24, Christ makes two observations about the days preceding his return: there would be intense tribulation as never experienced before,[1] and people would be living everyday life.[2] Those two scenarios seem contradictory, especially given that some religious organizations preach that there will be a one-world government birthing the Great Tribulation during that time. This teaching indicates that the Antichrist (a myth) will rule the government that will be so wicked that God has to destroy it and the whole world. According to this doctrine, God will miraculously remove (rapture) the people who have

accepted Christ as their Savior and Lord from this time of devastation. These religious organizations underpin this doctrine of the Rapture with Matthew 24:37-40 NASB:

> *"For the coming of the Son of Man will be just like the days of Noah. For as in those days before the flood they were eating and drinking, marrying and giving in marriage, until the day that Noah entered the ark, and they did not understand until the flood came and took them all away; so will the coming of the Son of Man be. Then there will be two men in the field; one will be taken and one will be left. Two women will be grinding at the mill; one will be taken and one will be left."*

The believers in the Rapture think the taking of one of the men and one of the women symbolizes God's snatching away those who have accepted Christ as their Savior and Lord. However, if we read these verses closely, the ones who were taken away on the day of the Great Flood were the ones who died, the ones who weren't in the ark with Noah and his family. Christ's point in this comparison and contrast of the day of his return to the day of the flood was that everyday life would be the same, but the death of people would be random rather than wholesale. This passage has nothing to do with a catching up of believers in a Rapture event.

(2) The emphasis on God playing favorites and giving up on the world isn't in line with His character. God is not a respecter of people,[3] nor does His love fail.[4] When Christ said, "*For God so loved the world ...*" in John 3:16 NASB, he clarified that our Father is not a hater; God would never leave human beings stranded in the dark energy of nothingness. So, despite the longings of some believers to escape from this "wicked" world, the Rapture is a myth. Furthermore, some of the same people who believe in the Rapture also believe that people who don't accept Christ as their Savior and Lord will spend eternity in Hell.

Again, sending people to Eternal Hell is totally out of character for God because He is not a respecter of people, nor does His love fail. This doctrine of Eternal Hell ignores the presence of nothingness in our world and our individual lives; it focuses on one action (accepting Christ) by each individual. This doctrine does nothing to change anyone on the inside. Some who believe in Eternal Hell label accepting Christ as being "born again" or "being saved." They seem not to understand the scope and the results of the new birth. What is essential to them is the one event of accepting Christ and escaping Eternal Hell.

(3) Just as the Rapture is a myth, Eternal Hell is also a myth. To understand why Eternal Hell is a myth, let's start with two verses. First Corinthians 15:28 NASB states that after the return of Christ is fully completed, God will be all in all. Hebrews 12:29 NASB says that God is a consuming fire. Remember,[a] God created an alternative to Himself to give His created beings a choice; at that point, He was no longer all in all. The culmination of Christ's mission (his return) will restore Infinity/Eternity to its original state through God's consuming fire. To understand how this will transpire, consider what happens when someone walks into a dark room and turns on the light. The instant the light appears, darkness disappears; we could say the light swallows up (consumes) darkness. Now, let's look at Isaiah 25:6-8 NASB:

> *The Lord of hosts will prepare a lavish banquet for all people on this mountain; ... And on this mountain He will swallow up the covering which is over all peoples, even the veil which is stretched over all nations. He will swallow up death for all time, and the Lord God will wipe tears away from all faces, and He will remove the reproach of His people from all the earth.*

This passage destroys the concept of Eternal Hell. Rather than blame the world's wickedness on the human race, God deals with its underlying cause, the covering that is over everyone. That covering is the deception that Lucifer uses to ensnare the human race. As God swallows up the deception, He swallows up death and Hell.

(4) As we have already discussed,[b] humans will put on immortality in two ways, determined by whether they are alive or dead when Christ returns. Since we're talking about the myth of Eternal Hell, let's look at the particulars surrounding the resurrection of the dead. First, Revelation 20:11-15 NASB focuses on what happens when God swallows up death:

> *Then I saw a great white throne and Him who sat upon it, from whose presence earth and heaven fled away, and no place was found for them. And I saw the dead, the great and the small, standing before the throne, and books were opened; and another book was opened, which is the book of life; and the dead were judged from the things which were written in the books, according to their deeds.*
>
> *And the sea gave up the dead which were in it, and death and Hades (KJV hell) gave up the dead which were in them, and they were judged, every one of them according to their own deeds. Then death and Hades were thrown into the lake of fire. This is the second death, the lake of fire. And if anyone's name was not found written in the book of life, he was thrown into the lake of fire.*

A surface read of this passage indicates that people are thrown into the lake of fire and that the lake of fire is Eternal Hell. But that's not it, not at all. If we pair the above verses with 1 Corinthians 3:11-15 NASB, we'll find the missing details:

> *For no man can lay a foundation other than the one which is laid, which is Jesus Christ. Now, if any man builds on the foundation with gold, silver, precious stones, wood, hay, straw, each man's work will*

> *become evident; for the day will show it because it is to be revealed with fire, and the fire itself will test the quality of each man's work.*
>
> *If any man's work which he has built on it remains, he will receive a reward. If any man's work is burned up, he will suffer loss; but he himself will be saved, yet so as through fire.*

The foundation that is Christ is "For God so loved the world" and "Love never fails." No other foundation exists. The lake of fire—God—isn't Eternal Hell. This point of judgment will be where God sets each person free from the works of deception and nothingness that keeps the individual captive to death.

(5) We might ask how fair is it to let people who have deliberately harmed other people escape the punishment of Eternal Hell. The answer is people will pay for their wickedness, just not eternally. The cleansing process of God's judgments at the Great White Throne Judgment will be fierce towards those who (whether they've accepted Christ or not) have harshly judged others or harmed others. Christ was clear on that in Matthew 7 when he admonished us that how we judge others is how we'll be judged[5] and that we should treat others the way we want them to treat us.[6] For some people, for those who have done great evil, this cleansing process will bring them to their knees as they fully understand and feel all of the pain and suffering they have caused to their fellow human beings. Then, after they comprehend the reality of God and the depth of their evil works, they will have a choice of life (immortality) or being consumed by God's fire (non-existence), as always. God isn't going to force anyone to choose immortality; however, when the human race sees and understands Him, everyone will more than likely choose life. We can have confidence in the power of God's life, light, love, and good.

(6) When we talk about what the Bible says about the return of Christ, we should keep in mind that the biblical writers had no concept

of today's world. The insights they—including Christ—gained from God were in terms of the context of the world they lived in and in terms they could understand. For example, in Matthew 24:15 NASB, Christ spoke of the *"Abomination of Desolation ... standing in the holy place"* happening before his return. He was speaking in terms of the Hebrew religion at that time. However, most religions have a sacred place where they worship their deity. The abomination in these holy places is the doctrines that misrepresent God and lead people away from the one-on-one relationship with God to a primary relationship with a community based on religious practices. In other words, the doctrines of the community outrank the reality of God. This abomination brings desolation because it causes people to cleave to the dogmas and religious practices rather than to God. The result is the desolation of the promise of death—people will die and either go to Heaven or Eternal Hell. The irony of this abomination in Christianity is that it celebrates Christ's resurrection as the pathway to Heaven, which we can only get to by dying. No, Christ's resurrection is the pathway to immortality, which comes into play at Christ's return and not before.[7] Death doesn't bring eternal life, even if one is "saved."

(7) The dogma of Heaven and Eternal Hell mimics the world's competitiveness—if you want to go to Heaven, join our team—that overtakes the world right before the return of Christ. Remember, the argument between Cain and Abel was about whose sacrifice was the best; that competitiveness in the world reaches a fever pitch during the transition between mortality and immortality. The images in Revelation 13, 17-18 take on a different meaning when we look at them in the context of competition:

- First, *the Beast from the Sea*[8] is the force of competition with the blasphemies it speaks, representing the relentless pressure for us to be better than other people (individuals, teams, groups, nations, etc.). This pressure shoves our focus from God to our fellow human beings and tries to invalidate His light, life, love, and good as being present in our world.

- Then, *the Beast from the Earth*[9] is the communication system that interconnects the world, with the sign of the fire coming down out of heaven being the radio waves that flow from earth to satellites to earth. The communications that flood our world feed the force of competition and push us to win at all costs.
- Next, *the Mark of the Beast*[10] could be the smartphones* that we hold in our hands or put to our ears. This mark brings an overabundance of images into our lives every minute of every day; most of these images are rooted in a competition where multitudes of people, manufacturers, businesses, organizations, etc., want our money, attention, and support.
- Finally, *the Number of the Beast*[11] (666) means that everything involved in these enemies against the human race is human-made. We built the competition (6), we developed and manufactured the communication system (6), and we created the smartphones (6)—there's nothing supernatural about them at all.

This relentless onslaught produces outcomes that impact the world and the individuals caught up by the competitiveness.

(8) Divisions among people diminish the likelihood of a one-world government. Remember,[c] the confusion of languages surrounding the Tower of Babel might have originated from competitiveness. How ironic is it that the writer of Revelation references Babylon (which means confusion) as part of the transition to immortality?[12] The constant assault of competitive messages and images creates tsunamis of chaos in our world. We hear conflicting news about everything: health, politics, money matters, education, retail, etc. This confusion even invades religious groups. Christ warned against acting on the conflicting messages in Matthew 24:23-28 NASB. In this passage, we could rephrase his words "*Wherever the corpse is, there the vultures will gather*"

to "Wherever the easy mark is, the scammers will gather." Religious people, among other groups, tend to be easy marks because they're trusting. (The way to avoid being scammed is to run to and stay in our prayer closets.)

(9) Perhaps the most severe consequence of this intense competition is the hardening of vicious circles where people learn to hate each other rather than love each other. The focus of any competition is to win (sometimes at all costs), which can denigrate the opposing side to a sub-human status. Since almost every individual and group wants to win, opposing sides often use identical words and actions to condemn each other. What is wrong-doing for our opponents is okay when we do it because the win is what is essential—this is hypocrisy at its worst. (Unfortunately, it's easier for people to see the other side's hypocrisy than to see their own.) Sometimes, the vicious circles that competition creates lead to intense actions that harm people who are not involved in the competition. The writer of Revelation gave a stark warning about the consequences of competition in Revelation 13:10 KJV: *"He that leadeth into captivity shall go into captivity: he that killeth with the sword must be killed with the sword. Here is the patience and the faith of the saints."* What many in the world call karma is a reality—going after the win is no excuse to rail against or harm those who aren't on an individual or group's team. The patience of our faith is to take heed to our words and actions; this can be challenging to do in a super-saturated atmosphere of competitiveness. Difficult, but not impossible. And as the world struggles with the pain of intense confusion fueled by this competition, the transition to immortality will begin. The two beasts, the mark, and Babylon will fail in their purpose.

Chapter Notes

[1] Matthew 24:21 NASB
[2] Matthew 24:37-38 NASB
[3] Romans 2:11 KJV
[4] 1 Corinthians 13:8 NASB
[5] Matthew 7:1-2 NASB
[6] Matthew 7:12 NASB
[7] 1 Corinthians 15:50-57 NASB
[8] Revelation 13:1-7 NASB
[9] Revelation 13:11-15 NASB
[10] Revelation 13:16-17 NASB
[11] Revelation 13:18 NASB
[12] Revelation 17:1-13 NASB

*We shouldn't panic about the smartphones because God will work them to our good if we ask Him to do so. Perhaps, with God's help, the phones will become tools that we use and not the rulers of our lives.

Remember

[a] Chapter 3, Paragraph 1
[b] Chapter 12, Paragraph 8
[c] Chapter 10, Mini-Essay

Mini-Essay

Biblical writers had no concept of what today's world would look like; even if they could visualize the details of our world, they wouldn't have had the vocabulary to describe it. So, the writers put the descriptions of the future world in the context of the world they lived in and the religion they understood. Since monarchies and emperors were the primary forms of human government in biblical times, that was the writers' context. The writer of Revelation defined Christ as the "King of Kings and Lord of Lords" in chapters 17 and 19. Some have seen the royalty description as evidence that Christ will return and ruthlessly govern the world. That's not it at all. Christ isn't returning to compete with his fellow human beings. As we'll see in the next chapter, what happens at that point is so much more than a coronation of the ultimate King. The return of Christ is when God's Somethingness completely swallows up nothingness. And God and Christ aren't the cause of that happening.

Let's take a look at the conversation Christ had with Pilate as recorded in John 18:33-38 NASB:

> *Therefore Pilate entered again into the Praetorium, and summoned Jesus and said unto Him, "Are You the King of the Jews?"*
>
> *Jesus answered, "Are you saying this of your own initiative, or did others tell you about Me?"*
>
> *Pilate answered, "I am not a Jew, am I? Your own nation and the chief priests delivered You to me; what have You done?"*
>
> *Jesus answered, "My kingdom is not of this world. If My kingdom were of this world, then my servants would be fighting so that I would not be handed over to the Jews; but as it is, My kingdom is not of this realm."*
>
> *Therefore Pilate said to Him, "So You are a king?"*
>
> *Jesus answered, "You say correctly that I am a king. For this I have been born, and for this I have come into the world, to testify to the truth. Everyone who is of the truth hears My voice."*
>
> *Pilate said unto Him, "What is truth?"*

Let's substitute the word "reality" for the word "truth." Christ's purpose was to testify to the reality of who God is and the choices we get to make. The kingship Christ claimed had nothing to do with other people (this world)–it had to do with his inner life. He had control of his destiny, and he knew that. At the same time, he listened to God's knowledge, understanding, and wisdom in making his choices. Christ never let go of God until he died; then, he was resurrected and expanded the immortality he was born with to the whole human race. Christ was king of his inner self, and that's what we're called to be.

Amid all the turmoil caused by the beast of competitiveness and the false prophet in the last days, the human race perseveres because of the reality of God's faithfulness. In Revelation 6:2 KJV, the writer relates what happened when Christ opened the first seal (the beginning of the action of His return): "*And, I saw, and behold a white horse: and he that sat on him had a bow; and a crown was given unto him: and he went forth conquering, and to conquer.*" The white horse represents righteousness—the righteousness of God that we become through Christ (2 Corinthians 5:21 KJV). This righteousness gives us the ability to conquer whatever keeps us from cleaving to God. We see white horses again in Revelation 19:11-16 KJV:

> *And I saw heaven opened and behold a white horse and he that sat upon him was called Faithful and True, and in righteousness he doth judge and make war.*
>
> *His eyes were as a flame of fire, and on his head were many crowns; and he had a name written, that no man knew, but he himself.*
>
> *And he was clothed with a vesture dipped in blood: and his name is called The Word of God.*
>
> *And the armies which were in heaven followed him upon white horses, clothed in fine linen, white and clean.*

> *And out of his mouth goeth a sharp sword, that with it he should smite the nations (deceptions): and he shall rule them with a rod of iron: and he treadeth the winepress of the fierceness and wrath of Almighty God.*
>
> *And he hath on his vesture and on his thigh a name written KING OF KINGS AND LORD OF LORDS.*

Along with the white horse representing righteousness, Christ also had a name that was his alone. In his words to the seven churches, recorded in Revelation 2-3, Christ mentioned the individual names several times. As we choose to cleave to God, we each transform into a new identity, a new name that only the individual and God know. The nothingness in our lives is swallowed up by God's Somethingness.

16

THE RETURN OF CHRIST

Key Words: Return, Christ, Eden, Lucifer, Eve, Deception, Knowledge, Good, Evil, God, Sound Waves, Destructive Interference, Adam, Appease, Love, Resurrection, Life, Religious Practices, Immortality, Heaven, Hell, Public Gatherings, Competitive, Faith, Doctrines, Prayer Closet, Reality, Reversal, Constructive Interference, Amplify, Vibrations, Universe, Confusion, Israel, Light, Good, Trouble, Eternal, Gospel, Judgments, Babylon, Consciousness, Resonance, Panpsychism, Synchronized, Human, Dissonance, Lamb's Marriage Feast, Infinity, Eternity, Virtual Choirs

The Model

(1) To understand what will happen at the return of Christ, we need to remember[a] the Garden of Eden and how Lucifer convinced Eve to eat the fruit of the tree of the knowledge of good and evil. He accomplished that feat by deception, as his voice (sound waves) destructively interfered with God's voice. While God honored Eve and Adam's choice of the knowledge of good and evil by not interposing Himself beyond what they allowed, Lucifer made sure the couple and their descendants experienced evil as he piled deception after deception upon

the human race. The drumbeat, "God has forsaken us, and we need to appease Him," was underneath all the deceptions, and the Earth soon seemed to reflect that rhythm. Centuries later, when Christ started his ministry, he appeared to be bringing in a new drumbeat of "God, our Father, loves us," until he intentionally gave up his life on the cross. At that point, Christ joined the drumbeat, "God has forsaken us." Lucifer's deception seemed to have won, and the human race appeared to be caught in that deception forever. But then, the resurrection happened, and the drumbeat of "God is our life" began to sound, only to be soon cast off because believers in Christ still died. As time went on, the early believers turned to religious practices and focused on the tenet that the way to appease God was to believe that Christ died for the sins of the human race. The concept of immortality gave way to the dogma of humans going to Heaven or Hell after death, along with the vague idea that Christ would return someday.

(2) Since the beginning of religion, the actions driving the efforts to appease God have centered on public gatherings and the doctrines of different faiths and sects. Through the years, some people have been so intent on promoting their religious practices and dogmas that countless competitive messages surround us today. Each group proclaims that their faith system is the way to God. Meanwhile, we have demoted Him to playing a secondary role as the religious practices and doctrines have taken center stage. Being faithful to one's religious group has become more imperative than cleaving to God. So, some of our relationships with God have become nothing more than telephone games (sometimes centuries-old telephone games) where one person told another person, who told another person, and on and on. How many of us believe what we believe because someone told us to believe what we believe? Like Job, some of us draw our concepts of God and the Bible from what other people have told us. And, for some of us, our prayer closets tend to sit empty while our public gatherings are full. Somehow—for Christ to return and immortality to manifest—

these circumstances have to change. Somehow, we humans need to understand the reality of what God desires for us, immortality.

(3) The key to the return of Christ is a reversal of what happened in the Garden of Eden when Lucifer slithered his way into Eve's world by falsely claiming to know the way for humans to be like God. The world needs constructive interference to amplify God's vibrations, which will counteract Lucifer's destructive interference. However, this need raises another question: Why doesn't God amplify His voice so that everyone can hear Him loud and clear? Again, the simple answer is God is love,[1] and love doesn't demand its way.[2] A relationship with God is available to all who desire it, but God will not push people into that relationship.[3] Cleaving to God is our choice to make, not His. Remember,[b] if the boundary between Infinity/Eternity and the Universe didn't lessen what God generates, His power would negate any choice for anyone. Since God won't destroy the border, the constructive interference needed to amplify His sound waves has to come from the human race. What we see in our world today is a lack of that constructive interference as the confusion of competition and its resulting deceptions swamp us. But, immediately preceding the return of Christ, that needed constructive interference begins and grows. The Book of Revelation describes this as two distinct groups of people amplifying God's voice until His light, life, love, and good fill the Universe.

(4) In Revelation 7:4-8 and 14:1-5 NASB, the writer describes the first group of people who amplify God's voice as a small number of individuals (symbolized as the twelve tribes of Israel)—individuals who never meet together as a group. These people are the ones who follow Christ wherever he goes and who begin to understand the voice of God. They start to sing a new song that only they can understand.[4] However, they aren't the only ones who verbalize the new things they learn. Revelation 7:9-17 NASB relates that multitudes of people join in to speak forth words amplifying God's voice. These individuals have

come out of great trouble, have discovered God and their prayer closets (where the cleansing happens), and have found the light, life, love, and good that God generates. As this happens, the new vibrations generate a "new" gospel, as described in Revelation 14:6-8 NASB:

> *And I saw another angel flying in mid-heaven, having an eternal gospel to preach to those who live on the Earth, and to every nation and tribe and tongue and people; and he said with a loud voice, "Fear God, and give Him glory, because the hour of His judgment has come; worship Him who made the heaven and the Earth and sea and springs of water.*
>
> *And, another angel, a second one, followed, saying, "Fallen, fallen is Babylon, the great, she who has made all the nations drink of the wine of the passion of her immorality."*

The new gospel is an eternal gospel based on God's judgments. Finally, people will distinguish between reality and deception because the clarifications between what is God and what is not God are the judgments of God. The influx of reality will change the choices people make. It will lead them to cleave to Him and put on immortality (the eternal gospel that has nothing to do with death). As more and more people cleave to God, confusion (Babylon) falls as the mysterious turns into the understanding of reality. Confusion's immorality turns into reality's immortality.

(5) Let's look at these circumstances from the scientific perspectives of consciousness and resonance, which will help our understanding. The concept of consciousness has been a continuing puzzle for scientists. While they've been able to test the physical responsiveness of the brain to stimuli, they've not been able to quantify the experience of consciousness—how do we measure our thoughts? Some scientists focus on the theory of panpsychism: all matter has its consciousness,

and together those consciousnesses constitute reality. This theory has some viability if:

- Throughout the creative process that produced matter, God's synchronized vibrations (words) were the foundation of that matter.
- Every iota of matter exists because of those synchronized vibrations and thus has consciousness.
- As some of the matter was formed into living organisms, consciousness levels grew more complex as the smaller units of matter resonated with each other in the new life.

God's synchronized vibrations pulled everything together to form a myriad of living organisms that resonated with Him. At the top of the living organisms were the human beings whose consciousnesses far surpassed what other life forms had. The humans had consciousnesses that found expression in the synchronized vibrations of language. As we have experienced, humans' language(s) directly impact what happens on our Earth because words generally come before actions. From this perspective, the human race rules over the Earth in a way that no other living organisms can rule.[5]

(6) From the time in the Garden of Eden when Lucifer destructively interfered with God's synchronized vibrations, the dissonance of deception has kept humans in confusion and at odds with each other. Some people want a just or peaceful world, but others see the world and its inhabitants as obstacles to overcome and subjugate. Even when the human race faces problems that affect everyone, we sometimes can't find solutions because of the dissonance of deception and the competition of various groups. The dissonance and competition have caused the world to become a humongous mess, as Christ foretold in Matthew 24 and the Book of Revelation described. Amid the turmoil, though, individual people find their places in their prayer closets and

begin to perceive and listen to God's vibrations. Their words and actions synchronize with God's beat of love, light, life, and good as they begin to understand reality. Their voices constructively interfere with God's sound waves which amplify His voice in the world. As the understanding of the reality of God grows, the confusion (Babylon) will fall apart. Here's what Revelation 19:1-2, 5-6 NASB says:

> *After these things I heard something like a loud voice of a great multitude in heaven (Infinity), saying, "Hallelujah! Salvation and glory and power belong to our God BECAUSE HIS JUDGMENTS ARE TRUE AND RIGHTEOUS, for He has judged the great harlot (Babylon) who was corrupting the earth with her immorality ..."*
>
> *And a voice came from the throne, saying, "Give praise to our God, all you His bond-servants, you who fear Him, the small and the great." Then I heard something like the voice of a great multitude and like the sound of many waters and like the sound of mighty peals of thunders, "Hallelujah, for the Lord Our God, the Almighty reigns!"*

[God's true and righteous judgments are the reality of His light, love, life, and good. Those judgments demolish the confusion of deception that has corrupted the Earth with darkness, hate, death, and evil.] In the above verses, we see the vibrations by the great multitude of the small and the great on Earth echoing the beats of Infinity. The statement "*Hallelujah, for the Lord Our God, the Almighty reigns!*" means that the reality of God's light, love, life, and good has overcome all the darkness, hate, death, and evil in the Universe.

(7) In Revelation 19:11-21 NASB, the writer describes what happens at the return of Christ with representations that seem to be autocratic, gory, and cannibalistic. Again, these illustrations are personifications of intangible realities: the kings of the Earth and their armies are the deceptions that have enslaved the world to evil, hate, darkness, and death. At the Lamb's marriage feast, those deceptions are picked apart and

consumed by the reality of God and His Christ (the Beast and the False Prophet are also consumed). Also, an interesting detail of this account comes in verse 11, *"And I saw heaven opened ..."*. Remember,[c] we talked about how God created the Universe to be an alternative to Himself with a definitive boundary between it and the rest of Infinity/Eternity. After God finished His basic Creation, three different levels of heaven existed from the perspective of earth-bound humans: the Earth's atmosphere (first), the Universe outside of the atmosphere (second), and the rest of Infinity/Eternity (third). Since the first and second heavens are not closed off from the view of the Earth, the heaven that opens has to be Infinity/Eternity. [Let's not forget that Christ is there with God.] Now the question is, how does Infinity/Eternity open up? God can't do it because He would be infringing on the choices that belonged to humans. Could the vibrations created by the voice of the great multitude of people cause the boundary between the Universe and Infinity/Eternity to disintegrate? That would be a reversal of what God did to create the alternative to Himself, but the reversal would not happen because of God's actions but because of the human race's words and actions. And as that reversal happens, the blobs of nothingness will be consumed by God's Somethingness as everything that God generates floods the Universe.

(8) Each of us has an individual part to play in the great multitude described above—a role that God has designed and will teach us one-on-one. The problem with many believers' religious practices and doctrines has been that they are all group projects where God is not the focus of the effort put forth. Dissonance has reigned in the world, with diverse groups clashing over who has the way to God. When we set aside the group projects and get one-on-one with God, we'll learn the music God has for each of us. As we are in our prayer closets, God will meld our voices together into the great choir that will bring Christ back. Our challenge is to connect with God apart from connecting with each other. As we begin to resonate with God's beat, deceptions will fall away, our words will change, and our actions towards others

will reflect all that God is. Immortality will become ours as God's love, light, life, and good swallow up hate, darkness, death, and evil.

Chapter Notes

[1] 1 John 4:8 KJV
[2] 1 Corinthians 13:5 KJV
[3] Matthew 7:7-14 KJV
[4] Revelation 14:2-3 NASB
[5] Genesis 1:26-28 NASB

Remember

[a] Chapter 7, Paragraph 2
[b] Chapter 4, Paragraph 4
[c] Chapter 3, Paragraph 2

Mini-Essay

Virtual choirs are a contemporary metaphor for how God will use humans' voices to accomplish His purposes. To work with these choirs, conductors make videos of themselves conducting with instrumental accompaniment and send them to the choir members. The choir members then create videos of themselves individually and send the recordings back to the conductor. The technicians turn the multitude of videos into a single video and release it. The performers are never in the same place, yet their voices blend into beautiful music. How God works with us is similar except for one significant difference. While the performers' interactions with the conductor are limited to the video of him conducting the choir, our interactions with God have no limits. He works with each of us in our prayer closets to help us perfect our contributions to the choir. And, as the number of choir members grows, so do the synchronized vibrations of their voices. As the writer of Revelation repeatedly points out, massive amounts of voices, sounds, and songs (too numerous to list) resonate as the return of Christ plays out.

Suggested Reading and Viewing

Constructive Interference | COSMOS.
https://astronomy.swin.edu.au/cosmos/C/Constructive+Interference

The largest virtual choir ever assembled. (2020, July 19). [Video].
https://www.youtube.com/watch?v=uS0ioMnVdHw

Voices, E. (2018, November 13). *Could consciousness come down to the way things vibrate? | Human World | EarthSky.*
https://earthsky.org/human-world/consciousness-what-is-theory-vibration/

17

OUR CALL TO ACTION

Key Words: Christ, Return, God, Enemies, Deception, Human, Kings, Babylon, Confusion, Nothingness, Elijah, Mount Carmel, Mount Horeb, Still Small Voice, Prayer Closet, Reality, Blessings, Cursings, Life, Death, Agency, Religious Practices, Public Gatherings, Knowledge, Wisdom, Immortality, Father, Lord's Prayer, Infinity, Holy, Lucifer, Vibrations, Forgive, Temptation, Evil, Kingdom, Power, Glory, Light, Love, Life, Good, Holy Spirit, Peace, Secret Place of the Most High

The Model

(1) Both the Old and New Testaments have numerous passages that predict what would be happening in the future, especially about Christ's return to the Earth. Most biblical writers understood this event as being great and terrible at the same time: great in that God "wins" and terrible in that the enemies of God are entirely defeated. These writers wrote in terms of those enemies being human because they (the writers) didn't grasp the relentless power of deception that stands between God and the human race. A way to look at it is that the human kings and leaders die, but the deceptions that keep people in bondage don't die. Remember,[a] it's not until the fall of Babylon, of confusion, that the deceptions are revealed and swallowed up by the reality of God and His purposes. Just imagine what it will feel like when we fully

understand the depth of deception trapping our world. If it weren't for also fully understanding the reality of God, we would be devastated about how much nothingness there has been in the world. The return of Christ presents another dichotomy: great joy for the reality of who God is and great remorse for our words and actions that were based on deceptions.

(2) One prophecy about the return of Christ in the Old Testament comes in its last verses, Malachi 4:5-6 NASB:

> *"Behold, I am going to send you Elijah the prophet before the coming of the great and terrible day of the Lord. He will restore the hearts of the fathers to their children, and the hearts of the children to their fathers, so that I will not come and smite the land with a curse."*

A surface read of this passage would bring to mind an end-time prophet who would do what Elijah did on Mount Carmel, where he exposed the powerlessness of Baal. In 1 Kings 18:20-19:14 KJV, the actual events of that interlude in Elijah's life went like this:

- He decimated the prophets of Baal after he revealed the power of the God of reality.
- Jezebel threatened to kill Elijah.
- He ran away from her to Mount Horeb and stayed in a cave for forty days before God's word came to him.
- At that point, a sequence of a great wind, an earthquake, and a fire happened before Elijah, but God was not in them. Then, a still small voice spoke to Elijah—the voice was God, who then talked with Elijah. [Interestingly enough, later on, Elijah was taken up into heaven without dying,[1] a possible foreshadowing of immortality.]

This last event was the epitome of the prayer closet, being alone with and talking with God. As we listen to God, He will expose the deceptions we believe and show us reality. Then, as the generations stop deceiving each other, their hearts will turn towards each other with the reality of love, light, life, and good. As everything God generates fills the Earth, His blessings will swallow up the curses that cause so much destruction. And life will swallow up death.[2]

(3) Perhaps the most crucial thing our prayer closets give us is agency. We are free to act independently and make our own choices within our secret places of prayer. So many religious practices and public gatherings are prescribed for us with words and concepts written by other people, often people who lived in the distant past. Not so with our prayer closets—they allow us to keep up-to-date with God's knowledge, understanding, and wisdom. Yet, some of us may hesitate to take advantage of our prayer closets simply because we don't know what to say to God. He is, after all, the Great Almighty who reigns in power and majesty. Getting up close and personal with Him can seem disrespectful, so we keep our distance and rely on our religious practices and public gatherings to make things right. Others look at the prayer closet as a place to go when everything else fails and we're in desperate need. We beg God to intervene. Still others see prayer as the method of bringing God's blessings to our lives; we couple prayer with conscientious giving to find prosperity and health. However, to find complete agency in our prayer closets and hone in on immortality, we need to push beyond the constraints of our religious beliefs and discover the reality of approaching God as His beloved children, who climb on His lap and talk everything over with Him.

(4) So, where do we start? Let's look at the passage where Christ talks about the prayer closet (KJV), the inner room (NASB) where we can be alone with God. Here's what Christ said in Matthew 6:5-9 NASB:

> *"When you pray, you are not to be like the hypocrites; for they love to stand and pray in the synagogues and on the street corners so that they may be seen by men. Truly I say to you, they have their reward in full, But you, when you pray, go into your inner room, close your door and pray to your Father who is in secret, and your Father who sees what is done in secret will reward you.*
>
> *"And when you are praying, do not use meaningless repetition as the Gentiles do, for they suppose that they will be heard for their many words. So do not be like them; for your Father knows what you need before you ask Him. Pray, then, this way ..."*

The presence of others in public prayers means our focus becomes split between them and God. Even with our eyes closed, our awareness of those around us never goes away, especially if people are praying aloud. Our divided attention positions us to not be in tune with God's small still voice because the ambiance of those around us can be louder than His voice. So, the place for us to start is to find settings where we can be alone. (If we find that challenging, we can ask God to "create" a prayer closet with us.) Then, we can choose to enter that solitude, close the door, and start talking. Let's look at that talking as pouring out our hearts to God rather than listing our prayer requests to Him. Remember, we are in a Father/child relationship, not in a Santa Claus/ Christmas list relationship. We are children of the Immortal God, not of a fabrication that lasts only a few months every year.

(5) Because trying to figure out how to approach our Father and talk to Him can be overwhelming for many of us, Christ gave us a template of topics that we can discuss with God. Ironically that template, found in Matthew 6:9-13 KJV, has been repeated as the Lord's Prayer countless times in religious gatherings. This repetitiveness has hindered us from personalizing the template as individuals in our prayer closets.

Let's look at each section and the questions we can ask to customize the concepts to our own individual lives:

- *"Our Father which art in heaven, Hallowed be Thy name."*
    - 1.) God is our Father. Like we all carry DNA from our birth parents, every human being has part of God inside them. Remember, matter is created through the energy, frequencies, vibrations, and nuclei of the elements that God generates; we are made from matter. Every one of us exists because God exists. He is, Christ is, and we are. Question to ask: *How does that work out in my life?*
    - 2.) God resides in Infinity/Eternity, the third heaven that Paul wrote about in 2 Corinthians 12:2-4 KJV. Remember,[b] the Earth's atmosphere is the first heaven, the Universe is the second heaven, and Infinity/Eternity is the third heaven. Since darkness exists in the first two heavens, we need God's light to see so we can put on immortality. Question: *How does God's light translate into my life?*
    - 3.) God is holy. God's holiness is in all He generates (life, light, love, and good) and through His words and actions. The Apostle Peter wrote in 1 Peter 1:15-16 NASB: *"... but like the Holy One who called you, be holy yourselves also in all your behavior; because it is written, 'YOU SHALL BE HOLY, FOR I AM HOLY.'"* Questions: *Am I generating what my Father generates: life, light, love, and good through my words and actions? Where am I missing it?*

- *"Thy kingdom come. Thy will be done in earth, as it is in heaven."*
    - 1.) God's kingdom is immortality—there is no death, darkness, hate, and evil in Him. Questions: *What part do I play in the coming of God's kingdom? How do I get to where I need to be for God's kingdom to manifest in me and then in the world?*

    - 2.) What is God's will? On the night of Christ's birth, the angels proclaimed (in Luke 2:14 KJV), *"Glory to God in the highest, and on earth peace, goodwill toward men."* Then, in John 3:16-17 KJV, Christ said that God sent him to save the world that He loved, and in John 10:10 KJV, Christ proclaimed that he came that the human race might have life more abundant than loss, death, and destruction. Question: *How can I work with God to make this happen?*

- *"Give us this day our daily bread."* In Matthew 4:3-4 KJV, when Lucifer tempted Christ to make stones turn into bread, Christ said, *"Man shall not live by bread alone, but by every word that proceedeth out of the mouth of God."* God's vibrations are a massive part of our very existence. They provide us with an enormous buffet daily: knowledge, understanding, and wisdom covering everything that is happening in our lives, our relationships with all the people in our lives, our concerns for our world, the deceptions in our lives, and on and on. The main buffet happens in our prayer closets with plenty of doggie bags to carry us through our daily lives. Questions: *What knowledge, understanding, and wisdom do I need today? Have I talked to God about what's puzzling me?*

- *"And forgive us our debts, as we forgive our debtors."*
    - 1.) Here is where we break out of the vicious circles that often trap us. Forgiveness happens when we don't demand personal revenge for another person's actions that offend or harm us. Deuteronomy 32:35 KJV tells us that vengeance and recompense belong to God—He knows how to deal with those who have hurt us. And He knows how to deal with us when we offend or harm others and how to bring about a clear understanding of our actions and the amends we should make. If we demand revenge, we open

ourselves to the vengeance of others when we offend or harm them. If we don't forgive, the vicious circles continue going around and around. Questions: *What do I need to comprehend about the situations where I've been in the wrong? How do I forgive people who have offended me? Father, how can You and I work together to bring good to everyone involved?*

- 2.) Forgiveness isn't forgetfulness. If others have offended or harmed us, we need to deal with the personal consequences of the offenses and hurts. If we try to bury what we're feeling, those emotions will still express themselves, often in mental or physical distress. If we turn to God to help us, He can work with us to find our places of healing and peace. Sometimes, the process can involve professional help, and we should never feel ashamed for asking God to guide us to the resources we need. Question: *How can I forgive when it hurts so much?*

- *"And lead us not into temptation, but deliver us from evil."* Our prayer closets allow us to be proactive in our approach to God rather than reactive. Our world can be a scary place. What makes us afraid, and what concerns do we have? By talking to God about those things, we can tap into His knowledge, understanding, and wisdom to avoid situations that make us victims of the deceptions we believe or victims of the harm others can do to us. Psalm 91 KJV centers on a powerful description of the benefits of dwelling in the secret place of the Most High, with protection and watch-care promised to the people who set their love upon God. Questions: *How do I deal with my fear about these things? What am I not seeing, Father? Is there anything I need to avoid? How can I stay under Your protection and watch care?*

- *"For Thine is the kingdom, and the power, and the glory, forever. Amen."* God's kingdom is what He generates: words and actions of light, love, life, and good. [He will never tell us to speak or act with darkness, hate, death, or evil.] God is who He is, and His Holy Spirit (His energy) undergirds everything He generates. The power of the Holy Spirit brought matter into existence and holds everything together. The glory of God comes as His Somethingness swallows up the nothingness of Lucifer's kingdom, first in our individual lives, then in our world, and then in the Universe. When we complete that process, we will put on immortality. Question: *What do I need to understand about Your kingdom and being Your child?*

While our prayer closets offer us an immense variety of topics of conversations with God, our prayer languages provide the foundation for the back-and-forth nature of those discussions. (Paul differentiated between praying with the spirit and praying with the mind in 1 Corinthians 14:14-15 NASB.) The purpose of the prayer languages is to bring God's vibrations into our personal spaces. Once the vibrations are there, we need to process the vibrations into knowledge, understanding, and wisdom. In other words, our brains need to turn the vibrations into meaningful communication so that we can pray effectual prayers. When we pray in our prayer languages, God speaks to us, and when we pray in our native tongues, we speak to God.

(6) Let's go back to one of the verses we looked at when considering how God created the alternative to Himself. In Isaiah 45:7 KJV, the prophet quoted God: *"I form the light, and create darkness: I make peace, and create evil: I the Lord do all these things."* We've already talked about the light, darkness, and evil, but we haven't pondered the reality that God makes peace. When we look at the world with all the competition, anger, and strife that divide the human race, achieving peace on Earth seems impossible. We gather ourselves into tribes that argue with each

other. Yet, if we look at our prayer closets as the venues where God makes peace, perhaps peace on Earth isn't such a stretch. With God as our private music instructor and as the conductor of the human choir, we each can learn the song He has for us to sing—the song of light, life, love, and good—and can synchronize with His beat. God knows how to meld all of our voices together in unity, which will swallow up the deceptions of nothingness and release the immortality waiting for us.

Chapter Notes

[1] 2 Kings 2:8-12 KJV
[2] Isaiah 25:8 KJV

Remember

[a] Chapter 16, Paragraph 4
[b] Chapter 3, Paragraph 2

Mini-Essay

Psalm 91 KJV is full of promises to those who dwell in the Secret Place of the Most High, who reside in their prayer closets. Since the world is such a dangerous place for us to live, let's read it together:

> *He that dwelleth in the secret place of the Most High shall abide under the shadow of the Almighty.*
>
> *I will say of the Lord, He is my refuge and my fortress: my God; in Him will I trust.*

Surely he shall deliver thee from the snare of the fowler, and from the noisome pestilence.

He shall cover thee with his feathers, and under his wings shall thou trust: his truth shall be thy shield and buckler.

Thou shalt not be afraid for the terror by night; nor for the arrow that flieth by day;

Nor for the pestilence that walketh in darkness; nor for the destruction that wasteth at noonday.

A thousand shall fall at thy side, and ten thousand at thy right hand; but it shall not come nigh thee.

Only with thine eyes shalt thou behold and see the reward of the wicked.

Because thou hast made the Lord, which is my refuge, even the Most High, thy habitation.

There shall no evil befall thee, neither shall any plague come nigh thy dwelling.

For he shall give his angels charge over thee, to keep thee in all thy ways.

They shall bear thee up in their hands, lest thou dash thy foot against a stone.

Thou shalt tread upon the lion and adder: the young lion and the dragon shalt thou trample under feet.

Because he hath set his love upon me, therefore will I deliver him: I will set him on high, because he hath known my name.

He shall call upon me, and I will answer him: I will be with him in trouble; I will deliver him, and honor him.

With long life will I satisfy him, and shew him my salvation.

[After Christ's resurrection, the promise of long life became the promise of immortality.] When we cleave to God, He will give us the knowledge, understanding, and wisdom we need to learn to overcome loss, death, and destruction. There are no magic wands, just our diligent work.

Is immortality a pipe dream? One way for us to test the concept out is to accept the call of action by finding our prayer closets and cleaving to God. What we discover in those secret places might be enough to change our world and bring us to immortality. It's worth a try.

NOTES

CPSIA information can be obtained
at www.ICGtesting.com
Printed in the USA
BVHW062017210622
640289BV00015B/1451

9 780578 354927